T. Blackburn

Further notes on Australian Coleoptera with descriptions of new genera and species

T. Blackburn

Further notes on Australian Coleoptera with descriptions of new genera and species

ISBN/EAN: 9783741183881

Manufactured in Europe, USA, Canada, Australia, Japa

Cover: Foto ©berggeist007 / pixelio.de

Manufactured and distributed by brebook publishing software
(www.brebook.com)

T. Blackburn

Further notes on Australian Coleoptera with descriptions of new genera and species

FURTHER NOTES ON AUSTRALIAN COLEOPTERA, WITH DESCRIPTIONS OF NEW GENERA AND SPECIES.

By the Rev. T. Blackburn, B.A.

The following notes and descriptions are founded chiefly on several collections made in the Northern Territory of S. Australia; I have included, however, among them the results of the examination of various Coleoptera from other parts of Australia that have recently come into my hands :—

CARABIDÆ.

HYPHARPAX.

H. DEYROLLEI, Cast.

In a note on this species in the Trans. Roy. Soc. South Australia, 1887 (p. 190), I drew attention to the anomalous characters of this insect, which seems to be very isolated among the Australian *Harpalides*, and also expressed a doubt whether I was acquainted with the male. I have recently procured on the sea coast near Adelaide, a male example which I have no doubt is conspecific with the females previously known to me, although its elytra and undersurface are darker in colour than any of them, and its antennæ have a little more tendency to infuscation towards the apex. A study of this specimen has satisfied me that the insect is much nearer to *Hypharpax* than to any other described genus; it presents differences however that may possibly indicate generic distinction, but as the sexual characters of species already attributed to *Hypharpax* present considerable diversity, I prefer for the present to regard *H. Deyrollei*, Cast., as a somewhat aberrant member of that genus. In the example before me the anterior tarsi are moderately dilated (scarcely less strongly than in typical *Hypharpax*), but the intermediate only very slightly; the hind femora are not toothed but they are somewhat dilated in a rounded manner at the place where the tooth is when present (it is quite likely that individuals vary in this respect); the hind tibiæ are strongly curved near the apex and their inner edge is moderately crenulate and fringed with long cilia.

CYCLOTHORAX.

C. PUNCTIPENNIS, Mcl.

This insect is extremely abundant all over South Australia; Mr. Macleay has done me the favour of confirming my identification of it. It is very close to *C. insularis*, Motsch., (of which I possess some specimens from New Zealand named by Mr. Bates), but may be distinguished from the latter by its narrower and more depressed form, and its less transverse protborax, which, moreover, is decidedly smaller in proportion to the elytra, while the rows of punctures on the elytra can scarcely be said to run in striæ. Capt. Broun in the "Manual of New Zealand Coleoptera," quotes Mr. Bates as stating that it (*C. insularis*) scarcely differs from the common Australia *Anchomenus ambiguus*, Er., the only difference observable being its more æneous colouring. I have not been able to find this statement in any of Mr. Bates' published memoirs to which I have referred, and possibly it may have been made in a private communication. Moreover, as Capt. Broun uses no inverted commas in his quotation, it seems doubtful whether he makes Mr. Bates responsible for the latter part of the statement. But not even Mr. Bates' authority (unless it were stated by himself to be founded on a comparison of the original types) could justify the statement. I have no doubt that *A. ambiguus*, Er., is a *Cyclothorax*, but Erichsen states it to be an insect with antennæ of pitchy colour the base being testaceous (whereas *C. insularis* has wholly testaceous antennæ, merely a little infuscate beyond the third joint), the elytra subæneous (the *absence* of which character, according to Capt. Broun's quotation of Mr. Bates, distinguishes it), the elytra half again as wide as the prothorax (which they are certainly not in *C. insularis*), and the systematic punctures on the elytral interstices placed quite differently (and very peculiarly) from those of *C. insularis*. But a further question arises whether *A. ambiguus*, Er., is the same insect as *C. punctipennis*, Macl., and this is not so easily answered. The only tangible differences seem to be that the antennæ are differently coloured, and the systematic punctures of the elytra differently placed. In *A. ambiguus* the position of the latter is described as so peculiar that it might well suggest the idea of abnormality. But the dark antennæ of *A. ambiguus* in a genus represented by many closely allied species, inclines me to the opinion that the identity of Mr. Macleay's species with Erichsen's wants confirmation, and I think that Mr. Masters has done wisely in retaining the two names,—for the present at any rate. The descriptions of the following new species of *Cyclothorax*, all from South Australia, points to the probability that *Cyclothorax* is largely represented on the continent.

C. OBSOLETUS, sp.nov.

Sat convexus; niger; antennis, palpis, pedibusque rufis; prothorace fortiter transverso, trans basin fortiter punctulato, lateribus rotundatis, angulis posticis subrotundatis minute subdentiformibus; elytrorum disco antice subtiliter quinquies punctulato. substriatis, striâ quintâ parte dimidiâ posticali obliteratâ.

[Long. 2½, lat. 1 line.

The head, antennæ, and palpi do not differ noticeably from those of *C. punctipennis*, Mcl. The prothorax is not much narrower than the elytra, and is nearly half again as wide as it is long down the middle, its base and front margin nearly equal, its sides very strongly rounded, the median line faint and abbreviated at both ends, the hind angles extremely obtuse but with a faint indication of being dentiform, the depressed basal area strongly, but not closely, punctured all across, a curved row of strong punctures running transversely a little behind the front margin.

Each elytron bears five rows of fine punctures placed in scarcely impressed striæ; of these the 1st stria is fairly well-defined and reaches the apex, but becomes impunctate in its apical half, the second is scarcely traceable to the apex, but its puncturation extends a little further back than that of the 1st, the next two resemble the 2nd, but their puncturation is a little more abbreviated, the 5th is scarcely defined or punctulate so far as to the middle of the elytron; under a Coddington lens a few punctures representing a 6th row are barely discernible; there is a strongly impressed stria a little before the margin bearing some strong punctures in its anterior half and about five large foveæ placed at equal distances apart in its posterior third; the marginal stria is punctured in its anterior third part. The colour varies somewhat, having a coppery tone in some examples with the middle part of the hind body and the prosternum inclining to red, and in some having the extreme lateral margins of the prothorax reddish.

A broader and more convex species than *C. punctipennis*, Mcl., with the prothorax much wider and more massive, the puncturation of the elytra evidently finer, and only five (instead of six) distinct rows of punctures on the same. From *C. ambiguus*, Er., it is distinguished by the colour of its antennæ, the much greater breadth of its prothorax, the two interstitial punctures of the elytra being both on the 3rd interstice, and probably by details of puncturation, but these latter are not indicated in Erichsen's description.

Port Lincoln.

C. FORTIS, sp.nov.

Convexus; ferrugineus vel piceo-ferrugineus; prothorace fortiter transverso, trans basin crasse punctulato, lateribus rotundatis postice rectis, angulis posticis acute rectis; elytris leviter 6-striatis, striis fere ad apicem sat fortiter punctulatis.

[Long. 2-2½, lat. ⅕ line (vix).

The head and its organs scarcely differ from those of the preceding species except in the antennæ being shorter and feebler. The prothorax is about half as wide again as it is long down the middle, its base evidently narrower than its front margin, its sides strongly rounded from the front nearly to the base, where they become quite straight and parallel, the median line faint and abbreviated at both ends, the hind angles sharply rectangular, the depressed basal area very coarsely (but not closely) punctured and longitudinally strigose, a strong unpunctured curved furrow running transversely a little behind the front margin. The sculpture of the elytra is very similar to that of *C. obsoletus*, except that there are six distinct (though lightly impressed) discal striæ on each, which are more strongly punctured, the punctures extending nearly to the apex except in the 6th stria in which they cease (or at least become very obscure) a little behind the middle, and that a 7th stria is faintly traceable like the 6th in *C. obsoletus*.

A considerably shorter insect than *C. punctipennis* and *obsoletus*, more strongly convex than either, and with the sides more rounded, the antennæ feebler and the thorax quite differently shaped. There is a slightly noticeable development of the apical external spine of the anterior tibiæ.

Near Port Lincoln ; also on Yorke's Peninsula.

C. CINCTIPENNIS, sp. nov.

Convexus ; piceo-rufescens; elytris piceis, marginibus lateralibus (late) et suturâ postice (anguste) testaceis ; antennis palpis pedibusque testaceis ; prothorace fortiter transverso, trans basin sat fortiter punctulato, lateribus rotundatis, angulis posticis subrotundatis minute subdentiformibus ; elytrorum disco antice fortius 5 punctulato-substriatis, tibiis anticis apice externa sat fortiter dilatatis.   [Long. 2½, lat. 1¼ lines.

A very robust species, more convex than *C. obsoletus*, with the anterior angles of the elytra considerably more prominent and the sides much more decidedly rounded ; there is very little difference in the head and prothorax except that the latter is somewhat wider in front, and the sculpture of the elytra scarcely differs except in that the punctures in the striæ are larger and stronger, are placed at wider intervals in the rows and scarcely exist behind the front half of the elytra. The evident sub-dentiform external prominence at the apex of the front tibiæ might almost suggest generic distinction were it not that a similar character is feebly displayed in *C. fortis*, which seems to be quite a typical *Cyclothorax* otherwise.

A single example in flood refuse on the banks of the Torrens.

C. PERYPHOIDES, sp.nov.

Minus convexus; niger; antennis palpisque rufescentibus; pedibus in parte ferrugineis; prothorace vix transverso, trans basin punctulato haud depresso, lateribus rotundatis postice rectis, angulis posticis acute obtusis; elytris vix striatis, 6-seriatim punctulatis, puncturis postice obliteratis.   [Long. 2⅔, lat. 1 line.

The head is somewhat narrower and more elongate than that of *C. obsoletus*, the antennæ and palpi as in that species. The prothorax is not much more than half as wide as the elytra, its length and width nearly equal, its base and front margin of nearly equal width, the sides very strongly rounded but becoming straight just before the base where it is sharply angled, but the sides of the base being somewhat oblique the angles are slightly obtuse; the median line is fairly marked but much abbreviated at both ends; the basal area is not depressed (as it is in all the preceding) but is similarly punctured, the punctures being considerably larger and more lightly impressed than in *C. obsoletus*; a strong unpunctured furrow runs transversely in a curve a little behind the front margin. On the elytra, the sutural stria is well-marked and attains the apex and is punctured in its anterior half; it can scarcely be said that there are any more striæ, but outside the sutural one there are five rows of punctures very similar to those of *C. cinctipennis*, and near the lateral margin the sculpture scarcely differs from the same in *C. obsoletus*; the apical third part of the disc is perfectly lævigate; as in all the preceding species of *Cyclothorax*, the abbreviated scutellar stria is indicated by a short row of punctures between the suture and the sutural stria. In my example the legs are black, except the following parts which are reddish—the anterior and intermediate coxæ and the underside of the corresponding femora, the extreme base of the hind femora and of all the tibiæ, and all the trochanters and tarsi.

In general appearance very much like a *Peryphus*. Differs from all the species of *Cyclothorax* described above in having the punctured basal area of the prothorax on the general level of the segment instead of being depressed.

Woodville, near Adelaide ; a single specimen.

DYTISCIDÆ.

CYBISTER.

C. GRANULATUS, Blackb.

Since the publication of my description of this insect I have seen a short series of both sexes. The peculiar sculpture of the elytra (which suggested the name) does not appear to be sexual, being quite as strong in the male as in the female, but it varies in both sexes,—some specimens showing it only feebly,—but it is always traceable. In the male the anterior tarsi are strongly transverse, the basal three joints together being considerably shorter than their width ; there is very little sexual pubescence on the intermediate tarsi, and the claws are rather strongly unequal.

LAMELLICORNES.

BOLBOCERAS.

B. SLOANEI, sp.nov.

Castaneum ; nitidum ; prothorace latera versus creberrime subtiliter rugulose, postice utrinque prope medium crebre crasse rugulose, in medio duplo (sparsius subtilissime et sparsim sub lineatim crasse), punctulato ; elytris punctulato-striatis ; striis suturam versus leviter, marginem lateralem versus fortiter, impressis ; pygidio crebre subtilius punctulato, dense hirsuto ; tibiis anticis externe 6-dentatis, dente basali parvo.

[Long. 10-11, lat. 6½-7 lines.

Maris capite cornu perlongo leviter recurvo (exemplo typico prothoraci longitudine fere æquali) instructo ; prothorace in medio fere ad basin late retuso ; parte retusa pernitida sparsim subtilissime punctulata, utrinque in medio cornuta, postice curvatim antice rotundatim utrinque profunde excavata.

Feminæ capite bituberculato; prothorace antice ad medium retuso, parte retusa (antice exceptâ) elevato-marginata.

The transverse carina at the base of the frontal horn of the male in front (*i.e.* the clypeal suture) is angulated in the middle, the horn itself being closely and rather finely rugose and simply (but not sharply) pointed at the apex and thinly clothed in its lower half with long fine hairs. The prothorax of the male is difficult to describe owing to the complexity of its sculpture; the lateral declivity (on either side) is very closely and rugosely punctured, finely in its lateral half, very coarsely in its middle half; this system of puncturation is continued narrowly and obscurely across the base and renders the portion of the surface where it prevails somewhat opaque; the whole of the remainder of the segment is extremely nitid, bears a system of very fine and very sparse puncturation, and forms (regarded as a whole) a great declivity, the surface of which is uneven in the following manner; its middle part (which is sulcate from the base half-way to the apex and bears a few large punctures) does not begin to be declivous close to the base, but runs forward a little distance as a flattened ridge on either side of which the declivity commences almost from the base itself, but in such fashion that its hinder edge here forms a curve on either side nearly touching the base in its middle, on its inner side margining the central non-declivous ridge (already mentioned), and externally forming a limit of the outside rugosely punctate surface, and then forming the hinder outline of a strong compressed horn which rises (on either side of the central declivity) about half-way between the base and apex of the prothorax, its height above the surface being about one-third that of the frontal horn; in front of each of the prothoracic horns the surface of the nitid declivity is disturbed (and its area extended laterally) by a deep round impression; the width of the space between the horns is considerably wider than the distance between the external base of either horn and the margin of the prothorax; the horns are inclined forward and upward. .

In the female the clypeus is strongly declivous its hinder edge forming a strong carina (most elevated in the middle);—from each end of the middle highly elevated part a strong carina runs obliquely towards the eye rising into a kind of tubercle at its apex where it is met by another carina given off from the extreme end of the carina that forms the hind margin of the clypeus; the back part of the head is elevated in a bifid tubercle. The prothorax is strongly declivous in its anterior part, the margin of the hinder part of the declivous space being prominent and conspicuous.

Mulwala, N.S. Wales; taken by Mr. Sloane, who has generously presented me with specimens of this and other interesting novelties.

B. CHELYUM, sp.nov.

Colore variabile, piceum vel piceo-ferrugineum (nonnullis exemplis elytris scutelloque læte ferrugineis); prothorace postice, ad latera crasse rugulose, in medio subtiliter, punctulato; post medium carina forti transversa (marginem lateralem haud attingente) instructo; scutello confertim subtilius punctulato; elytris sat fortiter punctulato-striatis; pygidio crebre fortius punctulato, dense hirsuto; tibiis anticis externe 7-dentatis, dente basali minuto vel obsoleto. [Long. 7-7½, lat. 4-4¼ lines.

Maris fronte antice cornu conico brevi (exemplo typico clypeo longitudine æquali) antice paulo inclinato, postice utrinque tuberculis acutis binis instructa; prothorace antice subperpendiculari, nitido, profunde sat anguste longitudinaliter excavato, excavationis lateribus utrinque in cornu acuto, capite vix breviori, productis.

Feminæ fronte antice bituberculata, postice tuberculis 6 transversim positis instructa; prothorace antice subperpendiculari, sat nitido, longitudinaliter leviter excavato, excavationis lateribus utrinque tuberculo conico instructis.

The long head horizontally projecting from the bottom of the almost perpendicular front face of the prothorax is very tortoise-like. The surface of the excavated part of the prothorax is almost lævigate in the male; in the female it is punctured rather more strongly but much more sparingly than the middle part of the disc behind the transverse carina, its puncturation consisting of large and small punctures intermingled. The prothoracic horns in the male spring from the sides of the excavation a little below the middle of their length and are directed almost straight forward.

This species resembles *B. laticorne*, Macl. The male may be distinguished *inter alia* by the *single* horn in the middle of the front of its head, the narrower and deeper excavation of its prothorax, and the much longer prothoracic horns which are pointed at the apex; the female differs in its clypeus less perpendicular, in the row of 6 tubercles (of which the point of the ocular canthus forms the external one on either side) being placed much more nearly in a transverse line, and in the declivous front part of the prothorax being more perpendicular and more sharply defined.

Mulwala, N.S. Wales; taken by Mr. Sloane.

MÆCHIDIUS.

M. SINUATICEPS, sp.nov.

Nigro-piceus; minus nitidus; minus convexus; sat parallelus; capite antice leviter bisinuato, lateribus obliquis fortiter bisinuatis; prothorace fortiter transverso crebre rugulose punctulato, transversim rugato, lateribus crenulatis sat fortiter arcuatis angulis omnibus acutis; elytris punctulato-striatis, interstitiis alternis leviter convexis; unguibus simplicibus.

[Long. 5½ (vix), lat. 2¾ lines.

The peculiar shape of the head seems to distinguish this insect from all others of the genus. The front margin is widely and very gently emarginate, but the emargination is distinctly (though gently) bisinuate. The sides of the clypeus are strongly bisinuate but in such fashion (their obliquity in front being comparatively slight) that the appearance of the clypeus bears a rough resemblance to that of a female *Liparetrus* of Mr. Macleay's first section (e. g. *L. phœnicopterus*, Germ.). The prothorax is not quite twice as wide as it is long down the middle; its sides are gently arched to behind the middle (where the segment is at its widest) and thence nearly straight (not at all sinuate) to the base; the front angles are decidedly, the hind very strongly, acute; the front margin is strongly emarginate, the base strongly bisinuate. Compared with the common *M. sordidus*, Boisd., the puncturation of the head is closer and not so strong, and that of the prothorax much better defined and running in transverse or oblique series so that the intervals appear as a system of transverse and oblique wrinkles. The sculpture of the elytra bears much resemblance to the same in *M. sordidus*. The setæ over the whole surface (at all events in the specimen before me) are not at all conspicuous. The anterior tibiæ bear three rather large very blunt external teeth. The underside is shining and coarsely and deeply punctured.

Northern Territory of South Australia.

LIPARETRUS.

L. LÆTICULUS, sp.nov.

Ovatus: nitidus; niger, antennis palpis pedibus et elytris (in parte) testaceis, pygidio rufescenti; clypeo antice truncato-lateribus obliquis; capite crebre rugulose, prothorace sparsim fortiter, elytris minus fortiter lineatim, propygidio (hoc per, magno) subtilius sat crebre, pygidio crebre fortiter, punctulatis; tibiis anticis externe bidentatis, antice in longo processu curvato productis; tarsorum posticorum articulo primo secundo duplo longiore; antennis 9-articulatis.   [Long. 1¾ lines.

An extremely distinct species not falling very naturally into any of Mr. Macleay's sections of the genus. Its clypeus bears much resemblance to that of *L. basalis* and its allies, but it has 9-jointed antennæ. The elytra are short scarcely reaching half-way from the base of the prothorax to the apex of the pygidium; they have no trace of geminate striæ and their sculpture consists of nearly regular rows of punctures; the testaceous colour occupies the whole surface except the base suture and lateral margins. The propygidium (in one sex at any rate) is enormous. The head, prothorax (except on the lateral margins) and elytra are glabrous, the propygidium, pygidium and underside sparingly furnished with rather short hairs. The two external teeth on the anterior tibiæ are small (the upper smallest) and sharp, the apical part of the

89

limb being produced almost in a spine curved outwards at the apex. Probably the other sex has the anterior tibiæ differently toothed, the elytra longer, and perhaps the pygidium and propygidium differently punctured.

A single specimen, sent by Mr. Rothe of Sedan.

L. SUAVIS, sp.nov.

Ovatus; minus nitidus; hirsutus; niger; pedibus et antennis in parte, palpis et elytris omnino, testaceis; clypeo antice subtruncato; capite crebre rugulose, prothorace vix evidenter, elytris crasse leviter, pygidio fortiter sat crebre, punctulatis; propygidio granulato; tibiis anticis (? alterutrius sexus solum) externe fortiter bidentatis; tarsis posticis gracillimis, articulo primo secundo vix longiori; antennis 9-articulatis.   [Long. 2¾ lines.

This species belongs to the same sub-section of the genus as *L. discipennis*, Guér., differing however from nearly all the other members thereof in having the elytra entirely testaceous with (at the most) a little infuscation round the scutellum. The head behind the clypeus is quite evenly convex; the anterior tibiæ are strongly bidentate externally; the hind tarsi are extremely short and slender, their basal two joints equal to each other in length. In other respects resembling *L. discipennis*.

Murray Bridge, &c.; in my collection, and in the South Australian Museum.

L. MYSTICUS, sp.nov.

Ovatus; nitidus; supra glaber, subtus sparsim breviter pilosus; ferrugineus capite (clypeo excepto) infuscato; clypeo antice rotundato-truncato vix emarginato, capite crebre subrugulose sat crasse, prothorace (huic lateribus ampliato-rotundatis) leviter sat crasse minus crebre, elytris sparsius irregulariter (striis geminatis nihilominus sat regulariter), pygidio propygidioque crebrius sat fortiter, punctulatis; tibiis anticis (? alterutrius sexus solum) externe 3-dentatis, dente primo minuto; tarsis posticis gracilibus, articulo primo secundo parum longiori; antennis 7-articulatis.   [Long. 2⅓ lines.

This insect has entirely the facies of an ordinary *Liparetrus*, but presents some structural peculiarities which might almost warrant the bestowal on it of a new generic name. Its antennæ having only seven joints will distinguish it specifically from all its hitherto described congeners, but the number of joints in the antennæ cannot be considered a generic character in the Australian *Heteronycidæ ;* the form of the anterior tibiæ (with the apical two external teeth very large and sharp, and a very small one above) and the very slight pilosity, are also exceptional. The distance from the apex of the elytra is very little less to the apex of the pygidium than to the base of the prothorax, so that a large piece of the propygidium is exposed, which (as also the pygidium) has no trace of a keel. The geminate striæ of the elytra are fairly well-defined ; the puncturation of the interstices similar to that in the geminate striæ, but not quite evenly dispersed. The prothorax is slightly more than twice as wide as its length down the middle, its base very little wider than its front, its sides very strongly and suddenly dilated and rounded in the middle, its hind angles quite rounded off, its disc distinctly channelled.

Taken by Mr. J. G. O. Tepper, at Monarto.

COLYMBOMORPHA.

There does not appear to me to be any sufficient reason for rejecting this name (as Mr. Masters has done in his Cat.). The structure of the claws and of the mesosternal process is very different from the same in *Calonota*. I can now add the information that the sexual characters are quite distinctive ; I received, some time ago, a short series collected in Western Australia by E. Meyrick, Esq., in which I find a single male (unknown to previous writers so far as I can ascertain). It has the antennal club very much longer than that of the female, and five-jointed.

DASYGNATHUS.

The species of this genus are very similar in appearance *inter se*, and unfortunately their published descriptions are not particularly

good, in no case I think instituting a comparison between one
species and another. *D. Couloni*, Burm., ought certainly to be
removed from the genus, and I propose for it the generic name
ADORYPHORUS. As I possess but a single example I am not in a
position, by dissection, to expose the generic characters fully,
but the character mentioned by Drs. Lacordaire and Burmeister,—
the atrophy of the upper lobe of the maxilla—together with its
small size and peculiar facies, render it an obvious error to
continue calling it a *Dasygnathus*.

According to all the hitherto published descriptions of the genus
the upper lobe of the maxillæ is devoid of teeth. I have recently
dissected a considerable number of specimens appertaining to it
and find that very few of them have this lobe toothless.

Up to the present time three Australian species that appear to be
rightly placed in *Dasygnathus* have been described, viz., *D. Dejeani*
♀, W. S. Macleay, *Australis* ♀, Boisd., *Mastersi*, (♂ & ♀) Macl.
The original type of the first of these is in the collection of the
Hon. W. Macleay, alongside which (Mr. Macleay tells me) is a male
Dasygnathus placed there (I understand) by the original describer.
Mr. Macleay has furnished me with a careful description of both
these specimens and has given me a male which he has compared
with the male just mentioned and found to be identical; he has
also favoured me with a detailed description of a male and female
Dasygnathus in the cabinet of Mr. W. S. Macleay labelled *D.
Australis*. With these materials before me, and also an assem-
blage of specimens of the genus from various collections, I have
prepared the following notes and descriptions of new species.

The specimens standing in Mr. Macleay's collection as *Dejeani*
and *Australis* must be regarded as representing those species
correctly. The following descriptions of them are compiled
(except that of ♂ *Dejeani* and ♀ *Australis*) from Mr. Macleay's
notes.

DASYGNATHUS DEJEANI, W. S. Macl.

♂. Blackish-pitchy, shining; the underside of a somewhat ferruginous tone and rather densely clothed with longish ferruginous

hairs except on the ventral segments where these hairs are concentrated in transverse lines (2 each on the basal 2 segments, 1
each on the rest) ; form very robust and gradually dilating almost
uniformly from the front to near the apex of the elytra ; clypeus
(with a very strong nearly erect reflexed margin) much narrowed
from base to apex, the front angles quite rounded off, the outline of
the sides and front gently convex. The forehead bears a very stout
recurved horn which is rather strongly punctulate to near the apex ;
the head behind the horn is impunctate or nearly so. The prothorax is just half again as wide as it is long down the middle, and
its base is just twice the width of its front ; the anterior angles
are well-produced but rounded at the apex, the hind angles obtuse,
the sides gently arched and not at all sinuate behind the middle ;
the anterior retuse portion extends backward to about the middle
of the segment, and nearly reaches the rugose lateral fovea on
either side, its hinder margin being strongly bi-tuberculate (the two
tubercles rather near to each other), and its surface very nitid and
punctured on the sides uniformly with the rest of the prothorax,
on the middle space more closely and confusedly (especially in
front where the sculpture is close and rugose) ; the rugose fovea
on either side short (not much longer than wide) ; the furrow
within the lateral margin is rugose, wide and deep ; that within
the anterior margin is very obscure in the middle and runs nearly
parallel to the anterior edge so that the space in front of it is not
much wider towards the middle than close to its ends ; the prothorax is not margined along its base which is broadly but not
deeply lobed in the middle with a foveate emargination on either
side. The elytra are at their widest considerably behind the
middle where their combined width is quite $\frac{4}{7}$ of their length down
the suture ; they are a little more than twice as long, and (together)
about a quarter again as wide, as the prothorax ; each of them
bears on the disc six well-defined punctulate striæ, of which the
first (close to the suture) attains the apex, the 2nd fails in the
apical fifth part, the rest are obsolete in about the apical third
part ; the interstices among these *striæ* are gently convex and are
impunctate except the front part of the interstices between the

1st and second and between the 5th and 6
or less puncturation ; the space outside and
striate area is rather finely and confusedly
are two fairly defined punctulate striæ i
lateral margin. The pygidium is densely p
with long erect hairs at the base and sides
its middle space is glabrous and much mo
The anterior tibiæ are strongly and sharp
external margin. The mentum is extrer
dinally concave behind and without lateral

The female (Mr. Macleay writes) has t
tulate, narrowed a little in front of the eye
with the anterior angles rounded ; the th
convex, apex moderately emarginate with
thickens into a small triangular extensio
median line, the sides rounded and more
the apex with strong punctures in the mar
angles slightly advanced, obtuse, the poste
the base rather wider than the apex, broad
in the middle with a foveate emargination
surface smooth and very minutely and thin
semi-circular, a little depressed and punct
scarcely wider than the thorax and nearly t
truncate at the base, scarcely widened behi
at the apex.

The remainder of the description is s
written above concerning the male.

excavation in front of the thorax smaller and more ci
scribed than in *Dejeani*, rugosely and finely punctate alou
median line, and surmounted by a transverse ridge which sca
shows any protuberances ; the rugose furrow near the side d
marked and quite one-third the length of the thorax ;
scutellum transverse, depressed, and punctate in the mi
the sculpture of the elytra much as usual in the genus
certainly smoother and more distinctly striate than in the fe
and more rugose at the apex ; the pygidium less pointed
more finely punctate than in the female, and not hirsute.*

[Long. 13, lat. 7

The following is a description of the female taken fr
specimen in my own collection :—

Reddish-pitchy, the head and prothorax darker, the und
clothed as in *D. Dejeani ;* form extremely robust, mode
dilated hindward ; clypeus broadly rounded with a strong n
erect reflexed margin ; the head evenly and rather c
rugulose, with the clypeal suture very little indicated exce
its middle, which is marked by a small well-defined tub
The prothorax is $\frac{2}{3}$ again as wide as its length down the m
and its base is decidedly more than twice as wide as its f
the anterior angles are rather strongly produced and some
sharp, the hind angles quite rounded off, the sides diverging
what straightly from the front to near the middle, and then g
arched (without any sinuation) to the base, which is ı
strongly lobed in the middle, is finely margined except i
middle half, and has an obscure foveate emargination on ı
side ; its surface is finely and thinly (most finely and thi
the middle of the disc) punctulate ; the furrow within the l

distant from the front of the prothorax nearly as far as the length
of the antennal club); there is a slight longitudinal concavity in
which the puncturation of the surface is at its strongest occupying the extreme front of the disc, a small slight indication (on
either side) of what in the male is the lateral fovea, and an
oblique impressed line on either side a little within the posterior
angle (this latter is possibly an individual aberration). The elytra
together are very nearly as wide as their length down the suture,
and are (behind the middle) a quarter again as wide as, but in
length not quite twice, the prothorax; their sculpture is very
much as in *D. Dejeani*, but the second, third, fourth, and fifth
striæ show a tendency to run in pairs. The pygidium is much
like that of *D. Dejeani*, but a little more pointed at the apex.
The external margin of the anterior tibiæ is cut into three very
blunt teeth.

N.B.—Mr. Macleay informs me that the ticket on his *D.
Australis* ♂ gives "*Scarabæus Juba*, Kirby," as a synonym, but
I hardly think it can be so. The only known Australian genus
presenting the cephalic and prothoracic characters ascribed to
S. Juba is *Corynophyllus* (*C. melas*, Fairm., agrees very well in
these respects); but the known species are much too small, and
it is improbable that Kirby could have omitted reference to the
antennæ if his insect had been a *Corynophyllus*. *S. Juba* should
be, I think, omitted from the Australian catalogue; or, better
still, relegated to an appendix.

D. TRITUBERCULATUS, sp.nov.

Robustus, postice dilatatus; nitidus; brunneo-piceus, capite
prothoraceque obscurioribus; subtus fulvo-hirsutus; clypeo antice
truncato vel obsolete emarginato, angulis rotundato-obtusis,
margine reflexo minus erecto; capite postice prothoraceque sparsim
subtiliter punctulatis; hoc leviter transverso, basi (lobo mediano
excepto) evidenter marginato; scutello fortiter transverso postice
obtuso; elytris punctulato-striatis, latera apicemque versus
confuse punctulatis, striis postice et ad latera obsoletis.

[Long. 12-12½, lat. 7 lines (vix).

Maris fronte cornu valido recurvo punctulato instructa ; prothorace antice excavatione sat parva (hac postice trituberculata,
tuberculo intermedio minuto) instructo. Segmento apicali ventrali
ut *D. majoris.*

Feminæ clypeo postice laminatim triangulariter producto, parte
producta super frontem inclinata; segmento apicali ventrali
utrinque punctis setiferis nonnullis instructo.

The prothorax is a quarter again as wide as long down the
middle, and evidently (but not much) less than twice as wide at
the base as in front. The male differs from that of *D. Dejeani*
as follows : the head is larger, the clypeus broader in front and
truncate, or even very slightly emarginate all across, with its
angles, though obtuse, not rounded off, and its sides slightly
concave in outline, its reflexed margin not quite so erect. The
prothorax is less emarginate in front, and is much less transverse, the furrow within the anterior margin considerably further
back and very much stronger leaving a wide space in front of it,
the anterior concavity very much smaller, not reaching back half-
way to the base, and occupying (transversely) much less than the
middle third of the segment, its concave surface almost impunctate except close to the front margin where it is only thinly
punctulate and not at all rugose ; the prominent hind margin of
the concavity is proportionally wider than in *Dejeani*, and has a
small tubercle in the middle ; the hind margin of the prothorax
is rather strongly margined except in the middle. The sculpture
of the elytra is on the same plan, but is altogether feebler and
less defined.

In the female the clypeus resembles that of the male in its shape
anteriorly, but its surface is closely rugose, and its hind margin
consists of two lines running obliquely backward and meeting in
the centre in an obtuse angle, the hinder part presenting the
appearance of a triangular lamina laid back (not quite flatly) on
the surface of the head, the apex of the triangle being the part
least closely applied to the head. On the prothorax the furrow
within the anterior margin is as in the male except that it is

continuous all across and in the middle is strongly and angularly produced backward in such fashion that the strip in front of it is narrow close to the anterior angles and dilates gradually towards the middle, but at the middle is suddenly and triangularly produced backward, till the apex of the triangle reaches back to a fifth of the length of the prothorax; there is a small punctulate fovea near the lateral margin on either side about half-way between the base and apex. The elytra resemble those of the male except that the sculpture is altogether stronger.

The anterior tibiæ are strongly and sharply tridentate on their external edge as in the male.

The mentum in both sexes is very rugose and hirsute, and strongly sulcate down the middle nearly to the front; in the male it is moderately tuberculate on either side. The upper lobe of the maxillæ is strongly dentate.

I have seen specimens from both N. S. Wales and Victoria.

D. MAJOR, sp.nov.

♂. Supra piceo-niger subtus fuscus fulvo-hirtus; sat nitidus; clypeo antice reflexo truncato, margine minus erecto; fronte cornu valido fortiter recurvo instructo; prothorace leviter transverso, (parte dimidia antica retusa postice fortiter bituberculata in medio longitudinaliter carinata, parte postica valde elevato-convexa supra depressa), inter partem retusam et marginem lateralem fovea profunda rugosa longitudinali instructo, mox intra marginem sulcato rugoso, sparsissime tenuissime punctulato, margine antico flavo-ciliato fortiter trisinuato; antice post marginem utrinque profunde sinuatim sulcato, margine basali integro; elytris sat fortiter punctulato-striatis, postice vix dilatatis, interstitiis nonnullis punctulatis; pygidio crebre (apicem versus minus crebre) subtiliter punctulatis; tibiis anticis obtuse tridentatis; segmento ventrali apicali apice fortiter arcuatim emarginato; mento utrinque fortiter tuberculato.

[Long. 13½-15, lat. 7½-8 lines.

♀. (? huj. spec.). Clypeo antice angustato minus reflexo rotundato-truncato, sutura clypeali carinata, in medio in tuberculo postice inclinato elevata; prothorace haud impresso, margine basali in medio interrupto, margine anteriori late (in medio triangulariter) dilatato, prope angulos anticos fovea rotundata instructo; tibiis anticis obtusissime bidentatis; segmento ventrali apicali haud emarginato, sparsim punctulato, punctis breviter setiferis.

[Long. 13½, lat. 7 lines (vix).

The horn of the male is very stout and strongly recurved, equal in length (on its front face) to the distance from the base of the prothoracic excavation to the front of the prothorax. The prothorax is about a quarter again as wide as long and somewhat less than twice as wide at its base as in front; its sides diverge very strongly for a short distance from the front and then are evenly and slightly arched to the base, the segment thus having a very quadrate appearance. On the elytra, the sutural stria reaches the apex; then comes a space bearing some strong punctures, irregularly placed, in front; then two punctured striæ nearly reaching the apex followed by another two much abbreviated; the interstices among all four punctureless or nearly so; between these striæ and the margin the elytra are scarcely striated but are finely punctulate, the punctures tending to run in rows, especially near the margin. The sinuous fovea on either side behind the front margin of the prothorax appears to be a narrowed and deepened continuation of the strong furrow that runs close within the lateral margin; it tends gradually away from the front margin and ceases abruptly about half-way to the central line of the prothorax. The carina running along the centre of the retuse part of the prothorax is feeble in front but becomes stronger behind, its hind apex being raised almost like a third tubercle. The large smooth round tubercle on either side of the mentum is a striking character. The superior lobe of the maxillæ has a strong tooth at the apex, and 3 or 4 smaller ones below.

The female which I have doubtfully connected with *D. major*, resembles it in its massive, yet moderately elongate form, scarcely dilated behind,—and in the sculpture of its elytra,—but the prothorax is less quadrate, and the front tibiæ furnished externally with only two (and those almost shapeless) teeth are very puzzling.

The sinuous fovea on the front of the prothorax is very similar to that of the male, but is continuous to the centre line where it is rather sharply angulated, the part in front of the furrow resembling in form a hood turned back quite across the front of the prothorax, extending back half a line in the middle, its centre part but little triangularly produced. There are no smooth tubercles on the mentum, but it is strongly convex with its posterior two-thirds deeply and widely sulcate in the middle. The male was sent to me by Mr. Sloane, and was taken near Melbourne. The female is, I believe, from N. S. Wales. Occurs also near Adelaide.

N.B.—A ♂ specimen (13½ lines) in the South Australian Museum, which I consider a variety of this insect, may possibly prove to represent a closely allied species; it has the clypeus narrower and slightly emarginate in front, and the prothorax less elevated behind with the lateral furrow continued from the sides all across the base (immediately within the margin as on the sides). It has the same elongate parallel form as my type of *D. major*.

Differs from *Dejeani* and *Australis* in the strongly dentate superior lobe of its maxillæ, and in the slightly elevated smooth linear keel that runs down the retuse portion of the prothorax in the male. The size of the prothoracic excavation varies somewhat but in all the examples I have seen it is exceptionally large. Both sexes differ from the two just named and from *trituberculatus* in their more parallel form, not dilated posteriorly; the male differs from that of *trituberculatus* by the presence of the keel on the prothoracic excavation, and the female differs from that of *Australis* by the sculpture of the head and clypeus, and by the different foveation of the prothorax.

D. RECTICORNIS, sp.nov.

Minus robustus, postice haud dilatatus; nitidus; brunneus vel rufo-brunneus, capite prothoraceque obscurioribus; subtus fulvo-hirsutus; clypeo antice truncato, angulis rotundato-obtusis, margine reflexo minus erecto; prothorace leviter transverso basi (lobo mediano excepto) evidenter marginato, sparsim subtiliter punctulato ; scutello subelongato postice subacuminato; elytris punctulato-striatis, latera apicemque versus confuse punctulatis, striis postice et ad latera obsoletis.

[Long. 12, lat. 6¼ lines.

Maris fronte cornu erecto sat gracili subtiliter punctulato instructa ; prothorace ut *D. trituberculati.*

Femina ut *D. trituberculati.*

Less robust in appearance than most of its congeners, and not dilated behind the middle of the elytra ; apart from this difference of form, and the totally different shape of the frontal horn in the male, I do not see any good character to distinguish this species from *D. trituberculatus.* The female before me has two round foveæ on either side of the prothorax some distance within the lateral margin—one near the front, the other behind the middle—but it is doubtful whether this can be relied on as a constant character.

The mentum of the male is smoothly tuberculate on either side, the cavity between the tubercles very deep and narrow (almost as in *D. major*) ; that of the female is very rugose and extremely deeply and widely convex almost from the apex. The superior lobe of the maxillæ is strongly dentate.

Taken by Mr. Sloane at Mulwala, N. S. Wales.

D. INERMIS, sp.nov.

♂. Sat elongatus, postice dilatatus ; rufo-piceus ; sat nitidus ; clypeo antice subelevato-emarginato : sutura clypeali sat elevata in medio laminatim angulatim elevata ; prothorace leviter transverso, antice leviter impresso, spatio concavo postice vix bituberculato, margine antico trisinuato, ad latera mox intra marginem sulcato-rugoso, antice post marginem anticam transversim sinuato-sulcato, (sulco in medio fortiter triangulariter retroducto), utrinque latera versus longitudinaliter rugoso-foveato, basi marginato ; elytris pygidioque ut *D. majoris* sculpturatis ;

tibiis anticis externe fortiter acute tridentatis ; segmento ventrali apicali apice sat fortiter arcuatim emarginato ; mento leviter convexo in medio leviter concavo.

[Long. 11, lat. 6 lines (vix).

· ♀ latet.

The absence of a frontal horn in the male at once distinguishes this species. The clypeus is shaped as in the females of the preceding three species, having its hinder margin formed of two oblique lines meeting in the middle in a sharp angle, the whole of this hind margin being laid back (as it were) on the surface of the hinder part of the head and being a little turned up to form the clypeal suture ; it is however more elevated (especially in the middle) than in any female known to me. The sulcus behind the anterior margin of the prothorax resembles that in the female of *D. affinis*, but is produced backward much more strongly in the middle, running down within the frontal impression on either side to its base (very little in front of the middle of the segment), where the two sides meet in a sharp angle. The prothorax is about a third again as wide as its length down the middle ; the whole segment, however, being smaller in proportion to the elytra than in any of the preceding species.

There is a single specimen in the South Australian Museum.

D. MASTERSI, Macl.

I have not seen a specimen of this insect, which is evidently very distinct from all its described congeners through the retuse portion of the prothorax in the male having a lateral protuberance on either side.

————

The above species appear to be perfectly distinct and separated by reliable characters. I have specimens before me which I believe to represent several other species, but they are closely allied to one or other of the preceding, and I am not sure without examining more specimens that their distinctive characters can be relied on. I find that in *Dasygnathus*, as in many other genera with strong sexual characters, these are liable to vary in their development. I have, in the case of each species, selected a well-developed male for description, but I have seen males of almost every one in which the characters are much enfeebled,—the tubercles on the mentum, and the size of the prothoracic excavation being particularly liable to variation. In many specimens the frontal horn is longitudinally concave on its anterior face, but this does not appear to be specific. The sculpture of the elytra is on the same plan (as described in the case of *D. Dejeani*) in all the species of *Dasygnathus* known to me, but varies in intensity so much within the limits of a single species that it would be misleading to characterise it particularly.

The following table will show the distinctive characters of the species :—

   A. Sides of the prothoracic excavation in the male devoid of lateral protuberances.

      B. Elytra conspicuously dilated to considerably behind the middle.

         C. Male with a recurved frontal horn.

            D. The prothoracic excavation more or less bituberculate behind.

               E. The prothoracic excavation rugosely and finely punctulate down the middle line (female with an anterior prothoracic impresssion) .... *Australis.*

              EE. The prothoracic excavation with the median line unmarked..... *Dejeani.*

            DD. The prothoracic excavation trituberculate behind.................. *trituberculatus.*

        CC. Male devoid of a frontal horn......... *inermis.*

      BB. Elytra not dilated behind the middle...

         C. Frontal horn of male recurved .... *major.*

         CC. Frontal horn erect....................... *recticornis.*

   AA. Sides of the prothoracic excavation in the male with lateral protuberances .... *Mastersi.*

ADORYPHORUS, gen.nov.

Mentum sat angustum, convexum, antice æqualiter angustatum, ligula vix distincta ; palpi labiales sub mente inserti, articulo ultimo ovali ; maxilla lobo superiore minuto cylindrico, apice penicillato ; mandibulæ haud dentatæ ; antennæ 10-articulatæ, flabello parvo ; tibiæ ut *Dasygnathi* formatæ ; tarsi posteriores modicæ, articulo primo sat elongato apice vix dilatato ; feminæ caput prothoraxque simplicia, hoc post marginem anticum ut *Dasygnathi* feminei (sed obsolete) sculpturato.

I propose this generic name for a small *Dynastid* which I have no doubt is identical with *Dasygnathus Couloni*, Burm., a species that certainly ought not to be associated with *Dasygnathus Dejeani*, W. S. Macleay. Unfortunately my specimen, like Dr. Burmeister's, is a female. I do not like founding a new genus without knowledge of the male, but as this insect has been described, and cannot, whatever its male may be, find a natural place in any hitherto characterised genus, I think I am taking the best course practicable in thus naming it. It agrees so well with Dr. Burmeister's description specifically that I need not add to that description beyond saying that in my example the colour of the upper surface is *pitchy* black rather than a genuine black, and the "small protuberance on the vertex of the head" is placed very far back and is very slight.

SEMANOPTERUS.

S. LONGICOLLIS, sp.nov.

Sat elongatus ; subparallelus ; convexus ; nitidus ; piceo-ferrugineus, rufo-hirtus ; capite transversim rugato, tuberculo conico armato ; prothorace canaliculato punctulato, tertia parte latiori quam longiori, lateribus postice haud sinuatis ; elytris sat fortiter minus oblique tricostatis, costis postice abbreviatis, interstitiis subtriseriatim punctulatis ; tarsorum posticorum articulo secundo primo multo breviore.   [Long. 7½-8, lat. 3⅜-3¾ lines.

Maris pygidio ad latera crebre, in medio sparsim, punctulato.

Feminæ pygidio crebre sat æqualiter punctulato.

Compared with *S. minor* this species is comparatively longer
and narrower; the head and prothorax are scarcely different
except in the greater length (in proportion to the width) of the
latter, and the absence of sinuation in the hinder part of its
lateral outline; the sculpture of the elytra does not run in quite
so oblique a direction; the sculpture of the pygidium is quite
different, as follows— in the male it consists of rather close
puncturation at the sides and very sparse in the middle, without
any transverse wrinkles, and in the female of close and almost
uniform puncturation with scarcely any trace of transverse
wrinkling; while in *S. minor* it consists of close puncturation in
both sexes (in the male, a little more sparse in the middle)
accompanied by a very conspicuous system of short curved
wrinkles or scratches; the second joint of the hind tarsi is barely
two-thirds the length of the first, while in *S. minor* the two are
about the same length.

From *S. angustatus* this insect may be distinguished by its
longer and narrower prothorax, and its pygidium only fringed
with hairs (while in *angustatus* fine erect hairs clothe the whole
of the surface), and from *S. convexiusculus* by its very differently
sculptured elytra. The other described species are all much
larger.

Coonabarabran, N.S. Wales; taken by Mr. Sloane.

S. MINOR, Blackb.

I have lately received from Mr. Sloane, of Mulwala, specimens
taken in various localities in N. S. Wales and Victoria which
have the sides of the prothorax behind much more strongly
sinuate (almost excised in fact) than in the type of this species,
but as I can discover no other difference whatever, and moreover
find some variability in this respect even in South Australian
examples, I do not think they can be treated as distinct.

PROTÆTIA.

P. MANDARINA, Weber.

This species, recorded from the Philippine Islands, is stated in the Trans. Ent. Soc. 1882, p. 156, to occur very plentifully in Queensland, and to be in the habit of attacking the hives of *Trigona* (the stingless bee) in great numbers.

Protætia is regarded by M. Lacordaire as a section of *Cetonia*. It seems singular, if the above statement is correct, that the insect has hitherto escaped the notice of Australian Coleopterists. I do not think that any of the species attributed to *Cetonia* in Mr. Masters' Catalogue are identical with *P. mandarina*.

BUPRESTIDÆ.

BUBASTES.

B. INCONSTANS, sp.nov.

Colore variabilis, cuprea vel ænea, vel viridis, latera versus plus minus cupreo-purpurea ; cylindrica ; capite sat fortiter minus crebre, prothorace (hujus latitudine majori basi posita) crebre minus fortiter, elytris crebre sat subtiliter, sat rugulose punctatis ; his apice obsolete emarginatis.

The head is slightly concave longitudinally with an impressed longitudinal line in the hinder part ; the eyes are sub-vertical, oblong, faintly sinuate on their inner margin, widely remote. The prothorax is nearly half again as wide as long, about half again as wide at the base as in front, its sides slightly converging and nearly straight from the base to near the front whence they converge more strongly and arcuately ; the front angles are obscure, the hind angles strong, acute, and pointed backwards, the base lightly bisinuate ; the true margin runs almost entirely on the underside (increasingly so from the hind-angles forward), and is quite obsolete near the front ; the surface bears an obscure

longitudinal furrow in its hinder half, which is deepened into a
fovea immediately in front of the base. The elytra are slightly
wider than the prothorax, are finely wrinkled in a transverse
direction in front, and bear a number of irregular feebly-defined
striæ which do not interrupt the general puncturation, and the
interstices between which are in places feebly convex and (in
most examples) here and there more or less lævigate. The underside is very sparingly clothed with fine adpressed hairs, and is
punctured more coarsely than the upper surface. The apical
ventral segment bears a very peculiar elongate transverse sulcus
in the hind face of its apex, which is much thickened; this
segment is in one sex truncate behind and level; in the other sex
it is turned up and rounded behind, and the penultimate segment
bears at the middle of its apex a short erect blunt spine.

B. LATICOLLIS, sp. nov.

Obscure æneo-cuprea, latera versus plus minus purpureo-
cuprascens capite prothoraceque (hujus latitudine majori in medio
posita) crebre confuse, elytris subtilius, subrugulose punctatis; his
apice oblique subemarginato-truncatis.

The head longitudinally is widely concave, but without any
distinct impressed line; it is rather coarsely punctulate in front,
the puncturation becoming finer and closer hindward. The prothorax is finely punctured on the anterior part of the disc and its
puncturation thence becomes coarser and less close hindward and
outward. [This sculpture of head and prothorax is of the same
kind as in the preceding species, but is evidently closer throughout.] The prothorax is about a quarter again as wide as long and
something less than half as wide again at the base as in front, its
sides nearly straight (but distinctly diverging) from the base to
about the middle whence they converge arcuately to the front;
the front angles (as in *B. inconstans*) are quite obscure owing to
the lateral margin being obsolete anteriorly, the hind angles strong,
acute and pointed backward; the surface bears a longitudinal furrow
feebly impressed, not reaching the front, rather deepened in front

of the base. The elytra are not in any part at all wider than the
widest part of the prothorax ; their sculpture scarcely differs from
the same in *B. inconstans* save in being slightly finer, the puncturation, moreover, being more evenly distributed and scarcely
interrupted on the interstices of the striæ which are less convex
in front and more so near the apex (but these last two characters
are slight and perhaps not very reliable) ; the outline of their
anterior margin is very markedly more strongly convex corresponding to the evidently stronger bisinuosity of the hind edge of
the prothorax ; their apical sculpture does not differ from that of
B. inconstans in any reliably specific manner, but the evenness of
the marginal outline is in average specimens even less interrupted,
while in some specimens there is an evident oblique truncation,
the extremities of which are defined though not spinose. The
underside and legs are conspicuously clothed with rather coarse
adpresssed short whitish scale-like hairs. The structure of the
apical ventral segment appears to be as in *B. inconstans,* to which
the present insect is closely allied, though differing considerably
in the shape of the prothorax, &c., &c.

The preceding two species of *Bubastes* both appear to be near
B. sphenoida, L. & G., so far as can be judged from the very brief
description of that insect in which scarcely any tangible characters
are mentioned, but " elytres bi-epineuses à l'extremité" will not
fit either of them. Moreover, there is a third species of *Bubastes*
in the Adelaide Public Museum in which the elytra are bi-spinose
at the apex, and which may be *sphenoida,* although I doubt it on
account of the puncturation being coarser than the description of
sphenoida would lead one to expect. From *Briseis conica,* L. & G.,
these insects differ in the non-denticulate margin of their elytra,
from *Eurybia* by their much stouter and more robust form, &c., &c.

ELATERIDÆ.

TETRALOBUS.

This genus presents extreme difficulty to the student, as far as
concerns its Australian species, owing partly to the close alliance
of some of its members to others, and partly to the insufficiency

of some of the earlier descriptions. I have lately had in my hands a considerable collection of examples taken in various parts of Southern Australia (from Eucla to Melbourne), and also from various parts of the Northern Territory, and have been unable to consider those from Southern Australia as representing more than one very variable species. It is extremly difficult to find two specimens absolutely identical. I find variation to an endless extent in the development of the furrows or foveæ on the head and prothorax, in the outline of the prothorax (especially in the degree of its dilatation about and in front of the middle, and in the degree to which its posterior angles are directed outward), in the distinctness of the striation and the puncturation of the striæ of the elytra, and in the shape of the apex of the same (some examples having them separately rounded with scarcely any trace of a mucro, some having them separately rounded with a distinct mucro, and some having them conjointly rounded with a more or less defined mucro).

Turning to the published descriptions, one finds *T. Australasiæ*, Gory, to be the original Australian species, to which, some years later, the Rev. F. W. Hope added *Manglesi* and *Fortnumi*. Between these latter, and between either of them and *Australasiæ*, there seems to be no really tangible distinction except size. Some years later M. Candèze added *M. Murrayi*, with the comment, " Very near *Australasiæ*, from which one will nevertheless distinguish it easily by the longitudinal furrow of the head and prothorax, and the much less strong pubescence." Regarding these distinctions I will observe that the latter is very likely to depend upon the freshness of the specimens, and that the former is sufficiently slight, because in the descriptions we find (*Australasiæ*) "fronte longitrorsum profunde sulcata, prothorace canaliculata," and (*Murrayi*) "front canaliculé et fovéolé, prothorax présentant une ligne lisse au milieu." A few pages further on M. Candèze says of *Manglesi* that it is very near *Murrayi* (although he judges from the description of the former that its head is more square and its elytra more distinctly punctulate-striate), and of *Fortnumi* that its distinction from *Australasiæ* is doubtful. A few years later still M. Candèze added another species from Southern Australia under the name *cylindriformis*, which he says must be placed beside *Murrayi*, and a comparison of the descriptions furnishes no tangible difference better than that in one the length and width of the prothorax are "subequal," while in the other that segment is longer than wide. Finally, in describing another species from Northern Australia (a very distinct one), he assigns it a place near *cylindriformis*, with a note that the latter species may be identical with *Fortnumi.*

My own impression is that all these five names represent one and the same species, and should stand in a catalogue as *Australasiæ*, Gory,—or at any rate the rest be relegated to an Appendix (which our Australian Catalogue sorely needs) of names not entitled without further evidence to a place in the body of the work.

The examples before me, which I consider as representing forms of *Australasiæ*, differ in length from 12 lines to 24 lines. The females are usually larger than the males and much more cylindrical with a decidedly stronger tendency to anterior dilatation of the prothorax. The head is more or less sulcate longitudinally, but the sulcus in many examples becomes feeble or even disappears before the front margin. The length of the prothorax down its middle is slightly more than its width across the base; the curve of its sides varies, being generally slight in the males and strong in the females in such fashion that in some examples of the latter the segment is wider just in front of the middle than its length down the middle; the disc is canaliculate (in some examples more strongly than in others), the channel usually abbreviated at both ends; the hind angles are sharp, more or less directed outward (most strongly so in the large females as a rule). The elytra are striated and the interstices are usually decidedly convex and closely and finely, but yet a little rugosely, punctured (the punctures a good deal run together by very fine transverse wrinkles); the striæ hardly distinctly punctured except near the shoulders and near the apex; in the largest females the interstices are usually less convex, and the puncturation of the striæ more evident. In one very large female before me the interstices are quite flat and the striæ punctured throughout as in ordinary examples they are punctured near the shoulders (this example, I am told, was taken in company with ordinary specimens). So far as I can judge, too, the females are less pubescent than the males, but this may be accidental. One of the females before me is exceptionally small, and resembles the male in the outline of its prothorax; it is just possible that it may represent a good species, but I cannot identify it with any described, and think it more probably a variety. It should be added that in all the specimens I have seen there are one or two vague impressions on either side of the prothorax near the lateral margin.

Monocrepidius.

M. Tepperi, sp.nov.

Fulvo castaneus; minus nitidus; minus elongatus; pube longiore sat dense vestitus; prothorace haud canaliculato, trans angulos posticos quam longitudine in medio latiori, a basi parum angustato, subtilius regulariter sat confertim punctulato, angulis posticis parum divaricatis bicarinatis; elytris prothorace angustioribus, a basi leviter attenuatis, apice vix emarginatis, fortius punctulato striatis, interstitiis sat planis leviter minus confertim punctulatis; prothoracis margine laterali antice in prosternum subducto; tarsorum articuli quarti lamella sat lata.

[Long. 5⅔, lat. 2 lines (vix).

The above mentioned characters would place this insect in the tabulation of species given by M. Candèze (Mon. II. pp. 191, &c.) in the same section as *Brucki* and *Jekeli*, the former of which is a very large broad species from Victoria, and the latter is a very anomalous insect (exact habitat unknown), of extremely parallel form with elytra twice and a half as long as the prothorax, while this is a very ordinary-looking *Monocrepidius*, with elytra of normal form and very evidently less than twice and a-half the length of the prothorax down the middle. None of the species described since (at any rate none from Northern Australia) are characterized as having the prothoracic margin passing to the underside and there forming the margin of a kind of prosternal gutter,—so I suppose it is distinct from them all.

Northern Territory of S. Australia.   Collected by Mr. J. P. Tepper, to whom 1 dedicate it, together with several other *Coleoptera* already described from the same locality.

M. JUVENIS, sp.nov.

Fuscus, antennis palpis pedibusque testaceo-flavis ; minus nitidus ; sat elongatus ; pube longiore sat dense vestitus ; prothorace haud canaliculato, trans angulos posticos quam longitudine in medio subangustiori, a basi evidenter angustato, subtilius regulariter crebre punctulato, angulis posticis haud divaricatis bicarinatis ; elytris prothorace vix angustioribus, a basi attenuatis, apice vix emarginatis, fortius punctulato-striatis, interstitiis sat planis leviter minus confertim punctulatis ; prothoracis margine laterali antice in prosternum subducto ; tarsorum articuli quarti lamella sat lata.   [Long. 5-6, lat. 1¾-1¾ lines.

This species is structurally very near to the preceding, and the sculpture and pubescence of its surface are very similar, but I think it certainly not a mere variety, as the entirely different colour is accompanied by very different proportions; nor are the differences sexual, as I have both sexes before me.   The present insect is a slender elongate form much narrowed before and behind; the other of robust appearance, rather short and parallel as compared with many of its congeners.   In *M. juvenis* the length of the prothorax down the middle is distinctly greater than its greatest width (across the basal angles), which latter, moreover, scarcely exceeds the greatest width of the elytra,—while in *M. Tepperi* the width across the basal angles of the prothorax is very evidently (about as 8 to 7) wider than the widest part of the elytra and than the length of the prothorax down the middle.   They differ, too, in respect of another character that I find not without its value in this difficult genus,—in *Tepperi* the external margin of the

prothorax is visible from above (that is on both sides from one
point of view) outside the external keel of the hind angle to
within a hairsbreadth almost of its hind apex,—while in *juvenis*
from a similar point of view, it seems to disappear under the
external keel considerably *before* its hind apex.

Northern Territory of S. Australia ; in my collection; also taken
by Mr. J. P. Tepper ; also by Prof. Tate.

M. PALMERSTONI, sp.nov.

Fusco-ferrugineus ; pedibus flavis, capite antice scutello pro-
thoracisque angulis posticis rufescentibus ; minus nitidus ; sat
elongatus ; pube longiore sat dense vestitus ; prothorace haud
canaliculato, trans angulos posticos quam longitudine in medio
vix latiori, a basi parum angustato, subtilius crebre subrugulose
punctulato, angulis posticis bicarinatis vix divaricatis ; elytris
prothorace angustioribus a basi attenuatis, apice rotundatis,
fortius punctulato-striatis, interstitiis sat planis leviter minus
confertim punctulatis ; prothoracis margine laterali antice in
prosternum subducto; tarsorum articuli quarti lamella minus
lata.   [Long. $3\frac{2}{3}$, lat. $1\frac{1}{2}$ lines (vix).

The narrower lamella on the 4th joint of the tarsi would
perhaps place this species in the last division of M. Candèze's
Section III., in which case its place in that section would be
beside *M. fictus,* from which the absence of a longitudinal carina
on the head will at once distinguish it. In M. Candèze's subdivision
of the earlier division it would fall side by side with the preceding
two species and the two others already mentioned. From the
latter two it differs as *M. Tepperi* does. From *M. Tepperi* (its
nearest real ally, I think) it differs by its very small size, somewhat
different coloration, less robust build, closer and slightly rugose
prothoracic puncturation, lateral margins of prothorax posteriorly
hidden by the external keel (as in *M. juvenis*), and apex of elytra
not at all emarginate (this latter may possibly not be a constant
character).

Northern Territory of S. Australia ; taken by Mr. J. P. Tepper.

M. FORTIS, sp.nov.

Fuscus, antennis palpis pedibusque pallidioribus ; minus elongatus, pube longiore sat dense vestitus ; prothorace fortiter
convexo, vix evidenter canaliculato, trans angulos posticos quam in
medio longitudine vix latiori, a basi vix angustato, confertim
subtilius subrugulose punctulato, angulis posticis bicarinatis vix
divaricatis ; elytris prothorace vix angustioribus, a basi parum
attenuatis, apice vix emarginatis, fortius punctulato-striatis, interstitiis planis crebre subtilius punctulatis ; prothoracis margine
laterali in prosternum subducto ; tarsorum articuli quarti lamella
sat lata.   [Long. 5⅓, lat. 1¾ lines (vix).

A species of robust build similar to that of *M. Tepperi*, and
belonging to the same group ; it may be distinguished from that
insect by its prothorax being more strongly and somewhat rugosely
punctulate with the disc strongly convex and the sides more
rounded, scarcely narrowed from the base to a little in front of the
middle, with its lateral margins much more hidden by the external
carina of the posterior angles which are less divaricate, and by the
interstices of the elytra having less tendency to convexity ; its
colour also is quite different.   From *M. Palmerstoni* it is distinguishable by its greatly superior size, different coloration,
longer hind angles of prothorax, &c.

Northern Territory of S. Australia ; collected by Prof. Tate.

M. VARIEGATUS, sp.nov.

Piceo-ferrugineus, capite, palpis, antennis, pedibus et elytrorum
dimidio basali, rufis ; robustus, sat latus, pube longiore sat dense
vestitus ; prothorace haud canaliculato, trans angulos posticos
quam in medio longitudine parum latiori, a basi angustato, crebre
sat fortiter subrugulose punctulato, angulis posticis bicarinatis
divaricatis ; elytris prothorace vix angustioribus, hoc duplo longioribus, a basi sat fortiter attenuatis, apice interno leviter emarginatis, striatis, striis antice fortiter postice gradatim obsoletius
punctulatis ; prothoracis margine laterali in prosternum subducto ;
tarsorum articuli quarti lamella sat lata.

Long. 6-8, lat. 1⅓-2⅔ lines.

An exceedingly distinct species. The proportion of colours on the elytra varies somewhat, the dark portion being sometimes limited to rather less than the apical half, in some examples occupying fully the apical half and even being produced up the suture a little beyond it; this dark portion seems to be always sharply defined and not gradually shaded off to the red part. The elytra are exceptionally short, very little exceeding twice the length of the prothorax. The lamella on the fourth tarsal joint runs out quite to the middle of the apical joint.

Northern Territory of South Australia; taken by Dr. Bovill.

HETERODERES.

H. CARINATUS, sp.nov.

Fusco-ferrugineus; antennis palpis pedibusque testaceis; sat elongatus, pube albida longiore sat dense vestitus; prothorace leviter canaliculato, trans angulos posticos quam in medio longitudine vix latiori, a basi parum angustato, confertim subtiliter et fortius sparsim punctulato, angulis posticis bicarinatis parum divaricatis; elytris prothorace vix angustioribus, a basi leviter attenuatis, apice rotundatis (intus vix mucronatis), fortius punctulato-striatis, interstitiis subplanis crebre subtiliter punctulatis; fronte longitudinaliter carinata; prothoracis margine laterali antice in prosternum subducto; tarsorum articuli quarti lamella minus lata. [Long. 5¾, lat. 1¾ lines.

Distinguished from *H. (Monocrepidius) albidus*, Macl., by the carina on its head.

Northern Territory of S. Australia; collected by Prof. Tate.

ACRONIOPUS.

A. PALLIDUS, sp.nov.

Rufo-testaceus, breviter pubescens; prothorace quam in medio longiore paulo latiore, hoc capiteque æqualiter crebre fortiter punctulato haud canaliculato; elytris punctulato-striatis, interstitiis convexis crebre subtilius punctulatis. [Long. 3, lat. ¾ line.

This insect has all the facies of an *Acroniopus*, and most of the structural characters—the convex forehead and front not margined, the antennæ with second and third joints very small, joints 4-10 sub-triangular and 11 elongate oval without appendage, the lateral margin of the prothorax not turned under at the apex, the posterior coxæ narrow and considerably dilated near the base but not dentate, the elongate basal joint of the posterior tarsi, &c., &c. ; but in some respects it seems to approach *Ascesis*, having the intermediate coxæ sub-contiguous, the prosternal sutures more curved than in *Acroniopus* (typical), and the fourth joint of the tarsi scarcely lamellated beneath. This joint has a small flattened space near the apex, but it does not seem to be a true lamella. The development of the lamellæ varies so much in some genera (*Monocrepidius* for example) that I do not like to found a new genus on this alone, and I think the best course is to refer the insect to *Acroniopus* with these qualifying remarks.

Northern Territory of S. Australia; taken by Mr. J. P. Tepper.

MALACODERMIDÆ.

TELEPHORUS.

T. TEPPERI, sp.nov.

♂. Rufo-testaceus ; elytris testaceo-brunneis pubescentibus, his apicem versus, antennis (basi excepta), genubus, tibiis tarsisque, infuscatis ; prothoracis lateribus pone medium concavis.

[Long. 4½ lines.

The prothorax is half again as wide as long, its front evenly convex, its sides gently curved to behind the middle, and thence dilated again to the base with which they form a sharp and prominent angle from which the base runs obliquely backward for a certain distance and then is slightly concave in the middle ; the basal and (especially) the lateral margins are rather widely and strongly reflexed ; the surface is shining and not punctured. The antennæ are more than half the length of the body, joint 1 equal to 2 and 3 together, 2 half the length of 3, 3-10 compressed

elongate triangular, 11 longer than 10, oblong and pointed, joints
1 and 2 wholly and 3 partially, testaceous.   The claws are simple.
Extremely like the European *T. fulvus*, Scop., except as regards
the differences involved in the above description ; the elytra, however, though much less shining are much more obscurely (scarcely
distinctly) punctulate.

Northern Territory of S. Australia ; collected by Mr. J. P.
Tepper.

N.B.—Mr. Tepper's collection contains two specimens (♂ and ♀)
of a *Telephorus* which I hesitate to distinguish from the above
specifically.   It is smaller (3-4 l.) and much more obscure in colour,
the parts characterized above as rufo-testaceous being pale fuscous,
with the sterna and basal portions of the ventral segments dark
brown, the tibiæ moreover being less noticeably darker than the
femora.   In the female the antennæ are only about half the length
of the body.

T. Palmerstoni, sp.nov.

Testaceus ; elytris pubescentibus, obscure cyaneis, margine
laterali antice flavo ; genubus tibiis tarsisque (plus minus) et
antennis (basi excepta) infuscatis ; elytris leviter rugulose punctulatis.   [Long. 3-3½ lines.

Apart from the entirely different coloration this species closely
resembles the preceding, but the following differences (though
slight) justify its being regarded as a distinct species,—the lateral
margins of the prothorax are only very slightly concave in outline
and its hind angles are very feebly marked ; the elytra are very
distinctly rugulose-punctulate,—almost as strongly as those of
T. fulvus, Scop.

Northern Territory of S. Australia ; collected by Mr. J. P.
Tepper.

The preceding two species appear to be genuine members of the
genus *Telephorus;* both appear to be distinguishable from such of
their congeners as bear a general resemblance to them by the
testaceous colour of the basal joints of their antennæ.

LAIUS.

L. VARIEGATUS, sp.nov.

Sparsim longe nigro-hirsutus; supra colore variegatus; capite cyaneo; prothorace fulvo, antice transversim late nigro-uni-lineato, disco maculatim infuscato; scutello cyaneo; elytris aureo-flavis, basi cyaneis, pone medium fascia versicolori (secundum marginem lateralem et suturam et circum apicem continuata) instructis; corpore subtus, antennis (articulis 3 basalibus flavis exceptis) femoribusque nigro-fuscis, tibiis tarsisque rufescentibus; capite subtilissime (antice crebre postice sparsim) punctulato; prothorace quam longiori tertia parte latiori vix evidenter punctulato; elytris crebre fortius rugulose punctulatis.   [Long. 2½, lat. 1⅓ lines.

On each elytron the base is entirely occupied by a cyaneous patch, which is extremely narrow at the suture and moderately so on the lateral margin, but on the disc runs down the elytron nearly a third of its length; a little behind the middle there commences a mark (of a reddish-brown colour variegated with cyaneous) the front margin of which is a curve extending from the lateral margin (about a third of the length of the elytra from the apex) to the suture very near its apex, and including the whole space to the apex except a large round spot of golden yellow colour similar to that of the middle part of the elytra. The basal joint of the antennæ in the specimen before me (I am doubtful of its sex) is elongate piriform, equal in length to the next two together; the 2nd springs from the external apex of the first (so that the antennæ appear to be geniculated in an external direction) and is longer than the next two together; the rest become gradually longer and more slender.

Northern Territory of S. Australia; collected by Mr. J. P. Tepper.

L. MAJOR, sp.nov.

Sparsim longe hirsutus; supra colore variegatus; capite nigro, antice cum labro rufo, inæquali, subtiliter sat crebre punctulato; prothorace rufo, quam in medio longiori quinta parte latiori, vix

evidenter punctulato, basin versus transversim impresso, lateribus
fortiter rotundatis, angulis posticis nullis, basi medio emarginata ;
scutello cyaneo-nigro ; elytris rufis, crebre sat fortiter subrugulose
punctulatis, magna parte humerali et fascia arcuata apicem versus
cyaneis ; prosterno abdomineque rufis ; meso- et meta-sternis
cyaneis ; pedibus (coxis, femoribus et tibiis anticis rufis exceptis)
nigricantibus ; antennis (articulis basalibus 2 rufis exceptis)
obscuris. [Long. 4, lat. 2 lines (vix.)

Maris antennarum articulis primo et secundo magnis ; hoc valde
depresso, intus valde angulatim dilatato, supra inæquali.

The hind margin of the humeral spot commences on the lateral
margin at a distance from its base of about a fifth of its whole
length, runs out in a curve (about three-quarters of the distance
across the elytron) towards the suture, and then proceeds obliquely
to the scutellum. The post-median fascia is in width about a fifth
of the length of the suture; in shape it bears a rough resemblance to
a horse shoe placed on each elytron with its convexity forward and
nearly attaining the middle of the elytron. The basal two joints
of the antennæ (in the male) are nearly equal to each other in
length, and together are quite as long as the head ; the 2nd joint
is attached to the external corner of the 1st ; on its inner side it
runs out from close in front of its base nearly at a right angle to
the line of the antenna, so that here the joint is as wide as
long, then with a sharp angle its inner outline runs sinuously
to the narrow apex of the joint ; on its outer side the joint is
gently curved ; of the remaining eight joints each is more slender
than the preceding one, the 1st and 3rd shorter than the other 5,
the 1st, 2nd and 3rd streaked with yellow, the apical the longest.

Differs from all other described Australian species of the genus,
inter alia, by its greatly superior size.

Northern Territory of S. Australia ; taken by Mr. J. P. Tepper.

NATALIS.

N. SEMICOSTATA, sp.nov.

Minus elongata; picea, nonnullis exemplis antennarum arto.
ulto. pedibusque rufescentibus ; illo valde compresso, superficie

compressa apice abrupte truncata interne acuminata ; elytris
antice crassissime, postice gradatim subtilius, cancellato-punctu-
latis, interstitiis alternis postice fortiter costatis.

[Long. 10-12, lat. 3-3½ lines.

The head has a small obscure depression between the eyes, and is
finely and closely punctulate, with some scattered punctures of larger
size.   The prothorax is slightly wider than long (in some examples
a little more so than in others) ; its sides are scarcely constricted
just behind the front and then a little rounded, so as to be at their
widest (in some examples scarcely so) a little in front of the base ;
its surface is punctured in the same fashion as the head and bears
(as usual in the genus) a longitudinal fovea on the disc, and an
angulate impression (not always strongly defined) near the front ;
its sides are strongly rugose.   The elytra to nearly the middle are
sculptured much as those of *N. porcata*, bearing longitudinal
lines the interstices between which are divided into quadrate
cavities by transverse lines, and the transverse being scarcely less
elevated than the longitudinal lines the latter appear scarcely
costate; but before the middle the transverse, and the alternate longitudinal, lines begin to fail, the latter soon disappearing,— so that
in the hinder part of the elytra the alternate interstices appear
as strong costæ bordered on each side with a row of fine punctures,
and having the intervals between them quite flat.   The shape of the
strongly compressed apical joint of the antennæ (having its flattened face abruptly truncate at the apex with one of the front
angles quite blunt and the other acute), seems to be distinctive of
the species.   In the examples before me (perhaps all of the same
sex) the two ventral segments preceding the apical two are closely
punctulate in the middle and densely clothed with golden pubescence, the rest of the ventral surface being sparsely and faintly
punctulate and thinly clothed with hairs.   There appears to be
some thin pubescence on the upper surface, but all the specimens
before me are evidently abraded.

Differs from the previously described species as follows, *inter
alia*,—from *Titana*, Thoms., in much smaller size, from *Mastersi*,
Macl., in the prothorax not being " much longer than wide,"—

from *porcata*, Fab., and *cribricollis*, Spin., in the shape of the apical joint of the antennæ.

Northern Territory of S. Australia ; in my collection, and taken by Mr. J. P. Tepper.

BOSTRICHIDÆ.

APATODES, gen.nov.

Gen. *Apaten* simulans, sed antennis clava lamellata terminatis.

I regret being unable to give the characters of this genus more fully, but unfortunately I have not a specimen before me in fit condition to bear the necessary manipulation. The resemblance to *Apate* is very close indeed, and as far as I have been able to investigate the structure it does not differ from that of *Apate* except in having the antennal club composed of three lamellæ, each of which is about equal in length to all the preceding joints taken together. This character alone is sufficient to justify generic separation. The basal joint is elongate, the 2nd very little longer than each of the next five, which are all very short.

A. MACLEAYI, sp.nov.

Nigro-brunneus ; capite (? alterutrius sexus solum) transversim 4-tuberculato ; prothorace parum transverso, antice ad latera spinoso, antice et postice obscure granulato, disco in medio sat rugulose tuberculato ; elytris crebre vix lineatim rugulosis, parte postica declivi, spinis 2 in medio instructa. [Long. 2¾, lat. 1½ lines.

The two spines on the elytra are placed half-way down the posterior declivity, one on either side of the suture, and point backward and outward.

This insect must very closely resemble *Bostrychus bispinosus*, Macl., (Trans. Ent. Soc. N.S.W. II. p. 276) and may possibly be identical with it, although apparently larger than that insect and scarcely fitting the description in respect of the prothoracic sculpture. But in any case *Bostrychus bispinosus* is a preoccupied name.

N. Territory of S. Australia ; collected by Mr. J. P. Tepper.

91

TENEBRIONIDÆ.

PLATYDEMA.

P. OBSCURA, sp.nov.

Ovalis : supra nigra, subtus picea, antennis, palpis pedibusque sordide testaceis; capite maris inter oculos cornubus 2 acuminatis antice directis instructo; prothorace quam longiori plus duplo latiori ; elytris punctulato-striatis, interstitiis punctulatis fortiter convexis.   [Long. 2¼, lat. 1⅓ lines (vix).

Resembles *P. tetraspilota* in shape. The prothorax is quite twice and a half as wide across the base as it is long down the middle and has its front margin truncate or nearly so, its hindmargin strongly bisinuate; it is margined all round, rather strongly and evenly narrowed from base to apex with its surface moderately and rather closely punctured, and an elongate fovea running forward from the base on either side about halfway between the middle and the lateral margin. The horns on the head are not much shorter than that segment and viewed from the side are triangular, their upper outline running almost straight forward. The eyes of the example before me are of a testaceous colour.

Resembles *P. critica*, Pasc., in which, however, *inter alia*, the prothorax is said to be less than twice as wide as long. Also probably resembles *P. Pascoei* and *laticolle*, Macl., (apparently described on females); it appears to be considerably larger than the former and differently coloured, and to differ from the latter *inter alia* by its prothorax strongly bisinuated at the base. The other described species are very different.

N. Territory of S. Australia ; collected by Prof. Tate.

TRIBOLIUM.

T. FERRUGINEUM, Fab.

This species is not included in Mr. Master's Catalogue; it is, however, plentiful,—doubtless introduced. I have it from South Australia and the Northern Territory.

Toxicum.

T. addendum, sp.nov.

Nigrum, minus nitidum, palpis tarsisque rufescentibus; prothorace antice elytris parum angustiori, postice angustato, fortius nec crebre (ad latera crebrius) punctulato, antice posticeque bisinuatis, lateribus sat rectis, angulis posticis subacutis; elytris parallelis, vix striatis, lineatim punctulatis, punctis sat validis, interstitiis haud punctulatis; antennarum clava 3-articulata; oculis haud divisis.

♂. Capite concavo, cornubus antice inclinatis,—anticis 2 parvis rectis acuminatis,—posticis 2 elongatis compressis apice hirsutis lateraliter æqualiter curvatis instructis.

♀. Capite haud cornuto, corpore fortius punctulato.

[Long. 5½, lat. 2⅓ lines.

The Australian species of *Toxicum* previously described having antennæ with three joints to the club and not differing much in size from the present species, are *distinctum*, Macl., and *parvicorne*, Macl. The former of these differs *inter alia* in the extremely strong puncturation of the elytra, and the parallel sides of the prothorax; the latter by the latter of the characters just mentioned and by the curve of the posterior horns being close to the apex.

In the present species the posterior horns are evenly bent inward from considerably below the middle, and are not far from meeting at the apex; the prothorax is more than a third again as wide as its length down the middle, and is at its widest immediately behind the front, whence it is very decidedly narrowed to the base.

N. Territory of S. Australia; collected by Mr. J. P. Tepper.

Hypaulax.

H. interioris, sp.nov.

Oblongus; convexus; supra minus nitidus; niger, antennis apice palpisque rufo-piceis; prothorace quam longiori quinta parte latiori, basi quam margine antico vix latiori, lateribus (pone medium leviter angulatim dilatatis postice sinuatis) incrassatis intus fortiter anguste sulcatis, basi incrassata haud bisinuata, angulis anticis rotundato-obtusis vix productis, angulis posticis parvis acutis extrorsum retrorsumque inclinatis, dorso nec foveato nec canaliculato ; elytris haud striatis, seriatim punctulatis, punctis modicis parvisque intermixtis, interstitiis planis minute coriaceis et punctulatis, basi late marginata ; mandibulis apice bifidis.

[Long. 9 lines, lat. $3\frac{1}{3}$ lines.

The mentum is moderately transverse, widely notched in front, finely punctulate, devoid of hairs ; gular furrow extremely strong and placed rather far back ; prosternal process preceded by a furrow (as in *H. Orcus*, Pasc.), its middle carina narrow and produced behind slightly beyond the lateral carinæ ; third and fourth ventral segments slightly sinuous behind ; epistomal suture fairly defined and arched ; labrum scarcely emarginate in front ; 3rd joint of antennæ decidedly longer than 4th ; scutellum very small.

The lateral margin of the prothorax is strongly thickened in its front third, and then is suddenly attenuated, thickening out again immediately in a kind of slight angular dilatation behind which it again becomes attenuated. The interstices of the elytra are perfectly flat except at the extreme base where they are very slightly convex ; there is no trace of the large punctures near the scutellum on the first interstice that are found in many species of the genus ; the shoulders are roundly obtuse and not prominent, the sides scarcely sinuous behind. The curvature of the anterior tibiæ is very slight.

The non-striate elytra marked with rows of mingled small and larger punctures, flat finely punctulate interstices, non-prominent shoulders, non-sinuate elytral apices, black legs, bifid mandibles, &c., taken together will distinguish this species from its congeners.

MacDonnell Ranges, Interior of Australia ; taken by Mr. A. S. Wild.

H. IRIDESCENS, sp.nov.

Oblongo-ovatus; sat nitidus, capite prothoraceque opacis sub-
iridescentibus exceptis; niger, antennis apice tarsisque rufescentibus; capite inter oculos bifoveato; prothorace quam longiori
dimidio, postice quam antice paullo, latiori; lateribus pone medium
dilatato-rotundatis, postice sinuatis, incrassatis, intus haud sulcatis;
basi vix bisinuata anguste marginata; angulis anticis rotundatis,
posticis acutis retrorsum inclinatis; dorso subtiliter obsolete longitudinaliter canaliculato; basi utrinque foveata; elytris sulcato-
punctulatis; punctis permagnis; interstitiis ad latera manifeste
nec fortiter, suturam versus vix, acute elevatis; basi minus crasse
marginata; mandibulis apice bifidis.　　　[Long. 8, lat. 3⅔ lines.

This species seems to oscillate between *Hypaulax* and *Chileone*,
which are, I think, too close to be treated as distinct. The
following characters have been omitted from the specific diagnosis
because if the two genera named above are to stand, this insect
might perhaps have to be treated as forming a third closely allied
genus. Mentum moderately transverse; angulated at the sides;
front margin notched in the middle; surface convex, subcarinate
down the middle with a depression on either side, hirsute (? only
in some examples); gular furrow moderate, placed well behind the
submentum. Epistomal suture well marked, curved. Labrum
rather decidedly emarginate in front. Joints 3 and 4 of antennæ
nearly equal, a little longer than the following joints; joints 8-11
gradually and not strongly thickened. Prosternal process distinctly turned up at the apex. Third and fourth ventral segments
sinuate behind. In other respects appears to agree with the
generic characters of *Hypaulax*.

It may be added that the punctures in the rows on the elytra
are placed far apart, and that there are about 12 to 15 punctures
in each row from the base to the beginning of the posterior
declivity (except the row nearest the suture which is bent round
in front nearly to the base of the third row, and so contains more
punctures); that the head and prothorax are scarcely visibly (or

very finely and sparingly in one example and that the front tibiæ are only gently cu

Resembles *H. opacula*, Bates, in many re seems to be quite different, the prothorax m the elytra very differently sculptured (in *op* striate" with their punctures "irregular, f run together"), &c., &c.

The elevated apex of the prosternal proce it bore a shining tubercle) is a notable comparative feebleness of the thickened elytra, and the evident (though slight) iride prothorax.

Northern Territory of S. Australia; Tepper.

LYGESTIRA.

L. SIMPLEX, Westw.

An example recently taken near Adelaic quite satisfactorily with the description c with that of *L. funerea*, Pasc., which, I sh tainly a synonym of the same species.

AMARYGMUS.

M. Blessig (Hor. Soc. Ent. Ross. 1861) *Chalcopterus* for certain species that attributed to *Amarygmus*, together with described insects, expresses a doubt whethe is to be found in Australia, stating, howev species he had been able to examine was

A. DIAPERIOIDES, sp.nov.

Ovalis ; niger, supra obscure cyaneus, epistomate labroque a
palpisque plus minus piceo-rufis, tarsis dilutioribus ; capite cr
prothorace sat sparsim, fortius punctulatis ; hoc basi quam el
rum basis vix angustiori, quam longitudo quamve margo au
dimidio latiori, lateribus leviter arcuatis, basi margineque a
bisinuatis ; elytris fortiter striatis, striis subtilius punctul
interstitiis leviter convexis sparsim subtiliter punctulatis ;
mentis ventralibus vix manifeste punctulatis, undatim longi
naliter strigosis ; antennis apicem versus manifeste incrassati

[Long. 2⅗, lat. 1⅗]

This is a true *Amarygmus* as distinguished from *Chalcopte*
it is very distinct from all the hitherto intelligibly desc
species, and does not seem to fit even any of Boisduval's la
diagnoses. It is perhaps nearest (but not very near) t
maurulus, Pasc.

Northern Territory of S. Australia ; collected by Mr.
Tepper.

CHALCOPTERUS (AMARYGMUS) AMETHYSTINUS, Fab.

This species belongs to the genus *Chalcopterus*, ha
mandibles truncate at the apex. It has been taken in
N. Territory of S. Australia by Mr. J. P. Tepper.
uniform bright blue colour (in some specimens with a v
tone in certain lights) of its upper surface and its red fe
together with its small prothorax, and elytra puncture
conspicuous rows (consisting of uniform rather strong punc
not placed very close one to another), the intervals bet
which are hardly visibly punctulate, render it an easily r

quam longitudo quamve margo anticus minus duplo latiori, lateribus pone medium subparallelis; elytris pone medium subdilatatis, fortiter striatis, striis crebre cancellato-punctulatis, interstitiis sat fortiter rotundato-elevatis sparsim subtilissime elevatis; segmentis ventralibus subtiliter minus crebre punctulatis, antice sat fortiter subreticulatim strigosis; antennis elongatis, apice vix dilatatis.   [Long. 8, lat. 3⅔ lines.

The elytra are more than four times as long, and (at their widest) quite half again as wide, as the prothorax; the nature of the puncturation of their striæ (arising from fine transverse carinæ connecting the raised interstices) is unusual in the genus.

N. Territory of S. Australia; taken by Mr. J. P. Tepper.

CISTELIDÆ.

METISTETE.

M. (ALLECULA) PIMELOIDES, Hope.

I have specimens (taken near Adelaide and in Kangaroo Island) of an insect which agrees very well with the description of this species except in respect of size. Mr. Hope gives 8 lines as the length, but the largest specimen before me does not exceed 7 lines; in allied species, however, I find so wide a variation in size that I do not consider this an important discrepancy. The insect is apparently a member of the genus *Metistete* (which, however, is very insufficiently characterised by its author). As the original description is very brief, I furnish a fuller one, as follows:—

Black; thinly clothed with erect hairs which are reddish towards the apex of the elytra; the front of the clypeus and of the labrum, the wide and conspicuous membranous connection between the 3rd and 4th, and the 4th and 5th ventral segments, the apex of the last ventral segment and the claws, red; coxæ more or less pitchy; antennæ obscure fuscous towards the apex, the apical two joints obscure ferruginous. Head little elongated, strongly but not coarsely punctured, the punctures very close and more or less (especially in the hinder part) running into each

other longitudinally; eyes large,—their distance apart equal
above to the length of the basal joint of the antennæ (below they
are very widely separated). Maxillary palpi with the 2nd joint
equal in length to the greatest width of the apical joint, which is
very strongly produced on the inner side so as to be transversely
triangular; the 3rd joint small and short but angularly produced
within. Antennæ equal to three-quarters of the body in length;
basal joint short and moderately stout, joint 2 very small, 3 quite
twice as long as 1 and 2 together, 4-6 successively shorter, the rest
not differing much in length but gradually a little more slender;
all the joints after the first rather slender. Mandibles broad and
slightly notched at the apex. Prothorax slightly wider than
long, slightly wider at base than in front, its front angles rounded
off, its hind angles slightly obtuse but well-defined, its sides rather
strongly rounded, its surface very convex (especially longitudinally)
and punctured uniformly with the hinder part of the head. The
elytra at their base are of the width of the base of the thorax;
they dilate gradually to a little behind the middle and then
contract to the apex, which is acuminate; the shoulders are quite
obsolete; each elytron bears 10 punctulate striæ of which the
first is abbreviated; the punctures in the striæ are somewhat
quadrate and very distinct in front, but become obsolete behind
the middle; the interstices are wide and flattish in front, becoming
gradually narrower and more convex hindward, and are transversely rugose and distinctly, but not very closely, punctured;
the epipleuræ are sub-vertical The scutellum is rather finely
and rather closely punctured. The legs are rather stout and very
long, the hind femora reaching nearly to the apex of the hind
body. The anterior tibiæ are angularly dilated within, just above
the middle in the ♂. The anterior four tarsi bear a lamella
under each joint except the last: of the hind tarsi the penultimate
joint only is lamellated; joints 2 and 3 together are on the front
tarsi slightly longer than, on the middle equal to, on the hind
shorter than, the first.

The apical ventral segment in the male is nearly twice as long
as the preceding segment; a forceps-like appendage projects beyond

it; each arm of the forceps is very wide, depressed and curved, so that the broad truncate apex of either is turned towards the other, and each angle of the truncate end bears a sharp hooked tooth; this appendage in many dried examples is only very partially exserted.

Judging from Mr. Newman's brief description of his *Tanychilus gibbicollis* the present species must be very near it, but seems to differ in its elytral striæ not being interrupted in front. If this interruption of the striation may have been an individual peculiarity of the type, it seems likely enough that *Allecula pimeloides*, Hope, (the insect here described as I believe) may be the same as *Tanychilus gibbicollis*, Newm.

M. (ALLECULA) ELONGATA, Macl.

The description of this insect points to its being congeneric with the preceding species and very close to it, but as there is no mention of the striæ on the elytra being punctured (other than the statement that the elytra generally are "densely and finely punctate,") I presume it is distinct.

M. LINDI, sp.nov.

Augusta; elongata; sat nitida; pilis erectis vestita; nigra, antennis pedibusque plus minus picescentibus; clypeo labroque antice, tarsis apicem versus, et abdominis segmentis apicalibus 3 postice, rufescentibus; capite crebre subfortiter, prothorace scutelloque sparsim minus fortiter, punctulatis; elytris striatis, striis (antice manifeste, postice vix perspicue) subtilius crebre punctulatis; interstitiis sparsim punctulatis, antice planis latis, postice convexis minus latis.

[♂ Long. 5, lat. $1\frac{3}{5}$ lines; ♀ Long. 6, lat. $2\frac{1}{5}$ lines.

♂. Tibiis anticis intus supra medium angulatim dilatatis; oculis sat approximatis; antennis elongatis; segmento ventrali apicali forcipite instructo.

♀. Tibiis simplicibus; oculis minus approximatis; abdominis apice haud forcipite instructo; antennis minus elongatis.

Very similar to the insect described above as *M. pimeloides*, Hope. Differs chiefly in the still narrower and more elongate form, in the very much less close puncturation of the prothorax, in the much smaller size of the punctures in the striæ on the elytra, and in the less convexity and more sparse puncturation of the interstices between the elytral striæ, which, moreover, are not transversely rugose.

The antennæ of the male are more than $\frac{3}{4}$, those of the female not much more than $\frac{1}{2}$, the length of the body. The forceps-like process at the apex of the hind body of the male is but little exserted in the single ♂ specimen before me, but it seems to resemble that of *M. pimeloides* except in the apices of the truncate ends of the forceps not being toothed,—but the specimen is so much damaged that possibly teeth may have been broken off.

The red colouring on the hind body is as in the preceding species. Port Lincoln.

APELLATUS.

A. PALPALIS, Macl.

An insect agreeing very well with the description of this species, and which I cannot doubt is identical, occurs all over S. Australia. During a recent visit to Port Augusta I observed it in the utmost profusion over the whole neighbourhood,—under bark of various trees, under stones, running on the ground, flying in the sunshine, and immolating itself in lamps at night. Individuals which I ascertained with certainty to be the females of this species agree perfectly with the description of *A. Mastersi*, Macl. The females, however, are very variable in colour and markings ; I have seen some examples agreeing in these respects with the males.

In the male the ante-penultimate joint of the maxillary palpi is very long and slender (scarcely shorter than the distance from the base of the antennæ to the apex of the labrum), the penultimate less than half as long and strongly dilated from base to apex, and the apical joint about twice the length of the penultimate, elongate-cultriform in shape with its outer margin strongly concave ; the antennæ are about half the length of the body, joints

1-3 moderately slender (2 very short, 3 a little longer than 1), 4 scarcely longer than 3, 5-10 shorter, 11 slightly the longest of all, 4-8 dilated (each more strongly in succession), 9 and 10 gradually less dilated, 11 slender ; the posterior tibiæ have a small tooth on their inner margin near the apex, and the eyes are almost contiguous on both surfaces of the head.

In the female the maxillary palpi are scarcely longer than the long joint in the male, the antennæ scarcely differ from those of the male except in the intermediate joints not being dilated, the posterior tibiæ are unarmed, and the eyes are a little more widely separated both above and below.

There are five ventral segments (of which the last is evenly rounded at its apical margin) in both sexes. The hind-body (except the base in some examples) is pitchy black.

The size varies from $2\frac{3}{4}$ to 4 lines.

A. APICALIS, sp.nov.

♀. Testacea, elytris abdomineque apice piceis ; capite prothoraceque subtiliter creberrime punctulatis ; elytris punctulatostriatis ; interstitiis (apicem versus convexis) subtilius sat crebre punctulatis.   [Long. 4, lat. $1\frac{1}{3}$ lines.

Extremely close to the corresponding sex of *A. palpalis*, Macl. Apart from colour and markings, the eyes are more approximate,— almost as close as in *palpalis* ♂,—and the head and prothorax are evidently more finely and closely punctured. The latter is also slightly less transverse, and more narrowed in front ; its width across the base is about a quarter again its length down the middle and very nearly twice the width of its front margin, the sides converge from base to apex with a very gentle curve, the front is nearly truncate, the base bisinuate, and there is an ill-defined wide impression down the hinder part of the middle between which and the lateral margin is a small basal impression on either side.

A single specimen was sent to me from Western Australia by E. Meyrick, Esq.

HOMOTRYSIS.

H. TRISTIS, Germ.

This species (on which the genus *Homotrysis* was founded by Mr. Pascoe) is extremely plentiful in South Australia. I feel no doubt that *Allecula carbonaria*, Germ., is identical with it. The author states that it is extremely close to *tristis*, but is a little larger, with the elytra not wider behind the middle and more deeply striated, and the prothorax more densely pilose. I have specimens, some larger and some smaller than average *tristis*, which display some or all of the other distinctive characters mentioned, but they do not appear to be specifically distinct. The characters of *Homotrysis*, as given by Mr. Pascoe, are very slight; one of them (viz., that the 2nd and 3rd joints of the anterior tarsi are "not longer" than the first) is very puzzling, as I do not know any *Allecula* in which they are longer, and in another sentence Mr. Pascoe speaks of the exceptionally short basal joint of the tarsi in *Homotrysis*.

H. (ALLECULA) FUSCIPENNIS, Blessig.

This is stated by its author to be near *A. carbonaria*, Germ., and is probably congeneric with that species. A comparison of M. Blessig's description of *A. fuscipennis* with Mr. Pascoe's of his *Homotrysis microderes* points strongly to the probability of their being identical specifically, in which case Mr. Pascoe's name must fall; both names were founded on specimens from Victoria. M. Blessig's descriptions. I may remark *en passant*, are models of lucidity, and his brief memoir on Australian *Heteromera* is in all respects admirable. Would that we all exhibited like ability and care!

CISTELA.

C. AUSTRALICA, sp.nov.

Ovalis; ferruginea; prothoracis lateribus et femoribus posticis obscure infuscatis: elytrorum lateribus (postice gradatim latius) et abdominis lateribus apiceque, nigro-piceis ; capite prothoraceque crebre sat fortiter nec rugulose punctulatis ; hoc transverso, semi-circulari, angulis anticis nullis, posticis acute rectis, basi late lobato (lobo postice emarginato), fovea parva utrinque ante basin posita ; elytris leviter punctulato-striatis, interstitiis sparsim subtilius punctulatis, manifeste transversim rugatis.

[Long. $3\frac{1}{5}$, lat. $1\frac{3}{5}$ lines.

The prothorax is almost a perfect semicircle, the base forming the chord ; at a casual glance the puncturation of its surface appears to be somewhat rugulose, but close examination shows that this is not the case. The blackish lateral margin of the elytra is very well defined ; at the base it is rather less than a third the width of the whole elytron, but it gradually dilates hindward till its inner margin meets the suture at a distance from the apex equal to about a quarter the length of the elytron, the whole apex thus being of a pitchy black colour.

This insect appears to be a genuine *Cistela*.

N. Territory of S. Australia ; taken by Mr. J. P. Tepper.

CURCULIONIDÆ.

MYLLOCERUS.

M. FASCIATUS, sp.nov.

Niger ; elytris squamis albis instructis, his fascias 2 formantibus (una basali, altera mediana), apice disperse albo-squamosis.

[Long. 2-$2\frac{1}{2}$ lines.

The basal two joints of the funiculus together are equal in length to the following five (which are subequal among themselves), the basal being a little longer than the second ; the scape nearly equals the whole funiculus, the club nearly equals the preceding four joints ; the antennæ are clothed with white hairs. The rostrum is wide and parallel. The prothorax is narrowed in front, is about half again as wide as it is long down the middle ; its sculpture is rugose, and a more or less distinct keel runs down the middle. The eyes are slightly oblong.

At once distinguishable from all the hitherto described Australian species of the genus by the conspicuous and well-defined elytral fasciæ formed of white scales.

N. Territory of S. Australia ; collected by Mr. J. P. Tepper,

M. DARWINI, sp.nov.

Piceus, squamis adpressis pallide viridibus (nonnullis piceis intermixtis) confertim vestitus; rostro brevi lato; antennarum funiculi articulo basali secundo parum longiori ; prothorace antice vix angustato, quam longiori dimidio latiori ; femoribus omnibus subtus dentatis. [Long. 2⅖ lines.

The uniformity and pale dead green colour of the scales on this insect (the intermixture of pitchy scales is noticeable only under a strong lens), together with its short broad rostrum, prothorax scarcely narrowed in front, and basal joint of funicle a little longer than the second, will distinguish this species from all its previously described Australian congeners.

N. Territory of S. Australia ; collected by Mr. J. P. Tepper.

LEPTOPS.

L. INSIGNIS, sp.nov.

Piceo-niger, elytris squamis fulvis albidis piceisque (maculatim et vittatim congestis) dense vestitis; rostro in medio acute carinato, vertice longitudinaliter subtiliter impresso ; prothorace crassissime rugoso ; corpore subtus pedibusque dense griseo-squamosis, his setis griseis vestitis. [Long. (rostr. incl.) 6-8, lat. 2⅕-3 lines.

In both the examples before me the head and prothorax are devoid of scales, possibly owing to abrasion, but the specimens appear to be very fresh in other respects. The latter, at its widest is very little more than half as wide as the widest part of the elytra ; it is slightly wider than down the middle it is long, its base truncate, its front margin rather strongly bisinuate. The elytra are punctulate-striate, the punctures in the striæ rather large, the interstices scarcely convex ; the whole surface is densely clothed with scales which form a sharply defined and intricate pattern.   The base is narrowly (somewhat more widely about the scutellum), pitchy; immediately behind it is a large transverse irregularly quadrate yellowish-fuscous patch common to both elytra and extending to the 6th stria on each (where it is at its narrowest) ; this is continued somewhat narrowly down the suture and a little before the hinder declivity spreads out again on either side, and here attains the 4th stria ; the scales on the lateral portions (which are much compressed) of the elytra (except in the front part) are greyish in colour, and this tint is widely continued round the apex ; the middle portion of the 5th interstice is quite white.   The elytra are much pointed at the apex, and the shoulders are laterally prominent in a subdentate fashion.   In one example before me several of the elytral interstices are a little costiform, but in the other example this character is absent.

The markings on the elytra resemble those of a *Stenocorynus*, but the strongly cavernous corbels seem to associate this insect rather with *Leptops*, from which I can discover no difference beyond the unusual character of the markings.

N. Territory of S. Australia ; taken by Mr. J. P. Tepper.

L. BAILEYI, sp.nov.

Oblongus ; niger ; plus minus sordide squamosus ; capite inter oculos et prothorace antice fortiter bituberculatis ; huic superficie tota tuberculatim rugosa ; elytris tuberculis magnis conicis et nonnullis minoribus 4-seriatim instructis ; interstitiis crasse rugulosis.   [Long. (rostro incl.) 7½, lat. 2½ lines.

The rostrum is about the length of the prothorax and is much dilated at its apex, the surface of which bears on either side a thick arched keel or crest ; the tubercles between the eyes are about the same size as the largest of those on the front half of the elytra, and are strongly compressed and longitudinally arched ; a very strong narrow central keel runs from a little behind the frontal tubercles nearly to the apex of the rostrum, but is interrupted between the tubercles; an obscure thick keel on either side connects the tubercles and the apical crests; the scrobes are flexuous and posteriorly obscure. The prothorax is about a quarter wider than long, flattened or slightly concave down the disc with two tubercles (about equal in size to those on the head) narrowly separated at the anterior margin; the whole surface is covered with small shining tubercles of unequal size; the sides are gently arched. The elytra at their base are scarcely wider than the prothorax and are widest about the middle; each elytron bears a sutural row of small tubercles, with a very large tubercle curved backwards at the summit of the declivity, followed by a row of five large tubercles at equal distances apart from base to near apex (the fourth the largest); then a row of four tubercles commencing behind the base, and finally two tubercles, one a little behind the shoulder, the other a little before the middle; the whole surface is coarsely rugulose and furnished with small obscure tubercles. The funiculus of the antennæ is very stout, the club nearly as long as the preceding four joints together, and (at its widest part) considerably wider than the funiculus (the joints of which are all subequal).

From all the previously described species of *Leptops* having interocular tubercles, this species appears to be well distinguished by the two large tubercles on the front of the prothorax. It is probably nearest to *L. musimon*, Pasc., which (besides the difference just mentioned) has the club of the antennæ not thicker than the funiculus, &c., &c.

Taken on Fraser Island and sent to me by F. M. Bailey, Esq., F.L.S., Colonial Botanist of Queensland, with whose name (so widely known among botanists) I have ventured to associate this insect.

L. FRONTALIS, sp.nov.

Ovatus, sat brevis; piceus, squamositate brunneo indutus; rostro unicarinato, scrobe lata postice obscura oculum haud attingente; capite in medio sulcato, inter oculos utrimque tuberculo

92

compresso instructo; prothorace fortiter transverso, in medio disci late impresso, rude vermiculato-rugoso; scutello vix manifesto; clytris prothorace fere duplo latiori, suturam versus obscure (marginem lateralem versus crasse profunde) seriatim punctulatis, singulatim tricostatis (costa interna postice tuberculis rotundatis consistente, externis subtuberculatis), humeris obliquis valde spinosis.   [Long. $4\frac{1}{2}$-$5\frac{1}{2}$, lat. $2\frac{2}{3}$-$2\frac{3}{4}$ lines.

Abraded specimens appear to be entirely black.   The lateral margins of the upper surface of the rostrum are thickened and convex, so that the rostrum might almost be considered tri-carinate; it is the upper apex of these lateral ridges of the rostrum which is raised into a compressed rounded tubercle immediately within each eye.   The frontal furrow is concealed beneath squamosity in fresh specimens.   The prothorax is nearly twice as wide as long down the middle; its sides diverge from the apex to near the middle, and then are almost straight to the base, and (owing to the extremely coarse vermiculate sculpture of the whole upper surface of the segment), they appear subtuberculate when viewed from above.   The distinctness of the tuberculation of the elytral costae varies, but I have not seen any example in which more than the costa nearest the suture (and that only in its hinder part) is distinctly broken into well-defined tubercles.   The shoulders resemble those of a *Catasarcus*.   The third joint of the tarsi is very little wider than the second.   The second ventral segment is equal to the following two together.

A very aberrant species of *Leptops*, but I can find no structural character of generic importance to separate it.

N. Territory of S. Australia; collected by Mr. J. P. Tepper.

Zymaus.

Z. (?) INCONSPICUUS, sp.nov.

Rotundato-ovatus; piceus, squamis brunneis et griseis dense vestitus (his ad latera, et prothoracis elytrorumque utrinque ad basin) vittatim congestis; rostro in medio late fortiter foveato,

fovea in medio carinata, scrobe curvata oculum haud attingente ;
capite in medio longitudinaliter impresso ; prothorace quam
longiori fere duplo latiori, leviter canaliculato, vermiculato-rugoso ;
scutello vix perspicuo ; elytris fortiter convexis, subrotundatis,
basi vix (in medio fere duplo) prothorace latioribus, obscure sat
crasse seriatim punctulatis, interstitiis subinterruptis minus convexis, horum nonnullis postice elevatioribus vix tuberculatis.

[Long. 3-4, lat. $1\frac{3}{5}$-$2\frac{2}{3}$ lines.

In a fresh specimen the sculpture is almost entirely buried under
the squamosity, which is of a dull brown colour except a wide
lateral vitta (indented three or four times within on the elytra),
and a short narrow vitta on either side of the middle common to
the prothorax and elytra, which are grey ; the squamosity of the
underside and legs is greyish rather than brown.   But only two
of the specimens before me are thus clothed, the rest being older
and more or less abraded, and in them the variegation of the
surface is not (or very little) noticeable.   In a very much abraded
specimen the rostrum appears tricarinate above (the lateral carinæ
being wide and feeble) and it is probable that this sculpture
always underlies the squamosity.   The eyes are very narrow,
vertical and acuminate beneath, the ocular lobes very strong.   The
triangular apical plate of the rostrum is strongly punctured and
concave down the middle.   This species has very much the *facies*
of a *Cneorhinus*.

The genus *Zymaus* is very briefly characterized by Mr. Pascoe,
as follows : " A *Leptope* differt unguiculis connatis." The present
species does not bear the slighest resemblance other than structural
to his species *(Z. binodosus)*, but as I can discover no other
structural character than that mentioned by Mr. Pascoe, to distinguish it from *Leptops*, I have no alternative but to call it by
the name *Zymaus*.

Northern Territory of S. Australia ; in my collection ; also
taken by Mr. J. P. Tepper.

LIPOTHYREA.

LIPOTHYREA (?) VARIABILIS, sp.nov.

Sat anguste ovalis (♂. ?) vel ovata (♀. ?); picea, squamis viridibus (super squamas cupreas positis) dense vestita; antennarum articulo secundo primo paulo longiore; capite rostroque plus minus distincte longitudinaliter subtiliter canaliculatis; prothorace quam longiori fere duplo latiori, antice angustato, (margine antico fortiter emarginato), postice truncato, in medio canaliculato, lateribus vix arcuatis; elytris postice abrupte declivibus, apice acuminatis (nonnullis exemplis subspinosis), punctulato substriatis, interstitiis 4° 7° et 10° rotundato-convexis.

[Long. 4½-6, lat. 2-2⅔ lines.

Freshly coloured specimens are uniformly and densely covered with bright green scales which appear to be very easily rubbed off, leaving the surface clothed with slightly shining obscure coppery scales, under which the derm is pitchy black; the legs, when denuded of scales, are of a more or less decided testaceous colour (especially the tibiae); in fresh specimens the sculpture is almost entirely buried under the scales.

This species presents the characters ascribed by Mr. Pascoe to his genus *Lipothyrea*, but appears to differ so much from the species he has described (*L. chloris*), that it is only with hesitation I assign it this place, and it is quite possible that it ought to be the type of a new genus of *Leptopsidæ*. The second joint of the antennal funicle being longer than the first is perhaps a generic character (certainly I think of greater importance in this group than in many), and it is not shared by *L. chloris*. The claws (Mr. Pascoe gives no information concerning those of *Lipothyrea*) are like those of *Leptops*, from which latter genus I hardly know how to separate the present insect structurally (though it differs much in *facies* from every *Leptops* known to me) except by the total disappearance of the *scutellum*. The rostral scrobes might seem to be distinctive, as also the shape of the rostrum itself, but *Leptops* varies in rostral characters.

Northern Territory of S. Australia ; collected by Mr. J. P. Tepper.

Oxyops.

O. INTERRUPTUS, sp.nov.

Minus brevis ; sat convexus ; niger, parce squamoso-setulosus ; rostro sat elongato, apice dilatato, medio postico carinato ; capite inter oculos fovea parva instructo ; prothorace quam longiori fere dimidio (quam margo anterior fere duplo) latiori, crasse confuse rugoso, medio et utrinque latera versus longitudinaliter depresso, disco pone medium carinato, a basi ad apicem arcuatim angustato ; scutello elongato elevato ; elytris sat elongatis, antice sub-parallelis, regulariter convexis, lineatim crasse punctulatis, spatia nonnulla rugulosa ferentibus, postice singulatim unituber-culatis, humeris externe conico-tuberculatis.

[Long. 7, lat. 3 lines.

The specimen before me (which may possibly be abraded) is thinly and irregularly clothed with small pale scale-like setæ. The sculpture of the prothorax consists of ridges or " wheals," among which are scattered coarse punctures, but the wheals are wanting in three vague longitudinal depressions, the middle one of which bears a carina in its hinder portion. The elytra are here and there strongly rugulose both between row and row of punctures and between puncture and puncture in each row, in such fashion that the non-rugulose portions appear as connected depressions forming on either side (a) a large lateral triangle (with its apex nearly touching the suture, and its base on the lateral margin, containing in its centre a little rugosity) in front of the middle; (b) a stripe running obliquely backward from about the second row of punctures to the lateral margin; (c) a vague space occupying the apical area in its half next the suture. The only tubercles on the elytra are a moderately conspicuous one near the apex of the fifth row of punctures and that on the shoulders, which is extremely conspicuous; it is, however, scarcely convex on its upper surface, but is directed outward, and has a slightly hooked appearance, though its apex is not sharp. The mesosternal projection is strong and sharp.

N. Territory of S. Australia ; taken by Mr. J. P. Tepper.

O. PARALLELUS, sp.nov.

Minus brevis ; subparallelus : fusco-ferrugineus, pedibus parum dilutioribus, albido squamoso-setulosus, setulis in elytris fasciam postmedianam formantibus; rostro sat elongato apice minus dilatato, in medio carinato ; capite inter oculos canaliculato, prothorace quam longiori quinta parte (quam margo anterior plus dimidio) latiori, crasse confuse rugoso, disco depresso in medio fortiter carinato ; scutello minus elongato, elevato ; elytris a basi postice leviter angustatis, sat convexis, fortiter cancellato-punctu-latis (interstitiis sat rugulosis), antice bituberculatis, interstitio 3° pone medium calloso.

[Long. 2, lat. 1⅓ lines.

Much less strongly narrowed behind than is usual in the genus.   In fresh specimens the hair-like white scales are con-densed upon the rostrum and the middle of the prothorax, on the scutellum, and especially on the elytra behind the middle, where they form a fascia very similar to that of *O. fasciatus*, Boisd. The sides of the prothorax are almost parallel from the base to the middle, where they are rounded, and whence they converge towards the front.   The base of each elytron is tumid from the humeral angle to near the scutellum, the extremities of the tumid region being more elevated than the rest (thus forming the two basal tubercles) ; the interstice on which the inner basal tubercle is situated is strongly carinate from a little before to a little behind the beginning of the apical declivity (thus forming the post median callosity), and several of the external interstices become somewhat carinate towards the apex, which consequently has a somewhat undefinedly uneven appearance.

The sculpture and markings of the elytra have a general resemblance to those of *O. fasciatus*, Boisd., compared with which this insect is of a different colour and much narrower and more parallel, the prothorax much more strongly carinate, with elytra more strongly foveate-punctulate and more strongly tumid near the base and more uneven behind; the mesosternal projection resembles the same in *O. fasciatus*. At a casual glance this species looks much like *Aterpus cultratus*, Fab.

N. Territory of S. Australia; collected by Mr. J. P. Tepper.

O. ARMATUS, sp.nov.

Minus latus, postice sat angustus; piceus, squamis griseis setulosis æqualiter (his nihilominus in scutello et longitudinaliter prothoracis in medio condensatis) minus sparsim vestitus; rostro sat brevi antice minus dilatato; prothorace quam longiori quintâ parte (quam margo anterior dimidio) latiori, sat fortiter ruguloso, postice utrinque leviter longitudinaliter impresso; scutello vix elevato; elytris sat convexis, fortiter seriatim punctulatis, interstitiis alternis leviter carinatis, quinta pone medium leviter tuberculata, humeris lateraliter acute spinosis; mesosterno antice acute producto. [Long. $3\frac{3}{5}$-$4\frac{2}{5}$, lat. $1\frac{2}{5}$-$1\frac{4}{5}$ lines.

A very distinct species, well characterized by its uniform grey appearance, with a whitish stripe down the prothorax and continued on the scutellum, while the elytra have no indication of tuberosity except in the fifth interstice being feebly callous behind the middle, and the shoulders having a strong sharp process directed outward.

N. Territory of S. Australia; collected by Mr. J. P. Tepper.

O. LATERITIUS, sp.nov.

Minus brevis; sat convexus; piceo-fuscus, interrupte parce breviter squamoso-setulosus; rostro brevi, lato; capite inter oculos profunde sulcato; prothorace quam longiori vix (quam margo anterior plus tertiâ parte) latiori, a basi ad apicem æqualiter angustato, basi bisinuato, æqualiter crebre subtilius punctulato; scutello elongato elevato; elytris a basi postice sat fortiter angustatis, striatis, striis crasse punctulatis, interstitiis punctulatis vix convexis, interstitio 3° basi calloso, lateribus sat longe pone basin

fortiter tuberculato ; femoribus apice fortiter incrassatis ; tibiis
omnibus intus fortiter denticulatis.   [Long. 3½, lat. 1¾ lines.

The arrangement of scales on the elytra is a good deal confused ; on each elytron there is an oblique fascia-like denuded
space immediately behind the middle, immediately in front of, and
behind, which the scales are at their greatest density ; but these
are in no part very conspicuous. The strong conical tubercle
close to the lateral margin of the elytra at about a fifth of their
whole length from the base, together with the strong (almost
angular) dilatation of the inner apex of the femora, and the strong
denticulations on the inner face of all the tibiæ, will render this
insect easily recognizable. The projection of the mesosternum is
obtuse and slight.

N. Territory of S. Australia ; taken by Mr. J. P. Tepper.

O. MODICUS, sp.nov.

Minus brevis; sat convexus ; piceus, antennis pedibusque rufescentibus; rostro brevi sat, lato ; capite inter oculos sulcato ;
prothorace quam longiori vix quarta parte (quam margo anterior
fere duplo) latiori, a basi ad apicem æqualiter subarcuatim angustato, sat fortiter minus crasse ruguloso, disco depresso in medio
carina forti antice abbreviata instructo ; scutello sat elongato
elevato ; elytris striatis, striis crasse fortiter, interstitiis crebre
subtilius, punctulatis, his alternis antice convexioribus, humeris
externe obsolete prominentibus.   [Long. 3, lat. 1⅔ lines.

The scales on the head are a little condensed, and rather
elongate between the eyes ; those on the prothorax are evenly
distributed and sparse ; those on the elytra are much more dense
(especially in the apical half), and more or less conceal the sculpture except on a space (more or less interrupted by squamosity)
commencing immediately behind the anterior declivity, extending thence backward to about the middle of the elytra and
limited laterally by the suture and about the 6th interstice (this
is very likely to be the normal state of the insect, as I have two
specimens before me thus clothed). The shoulders show a

decided tendency to prominence in a lateral direction,—but cannot be called "tuberculate."

A very obscure-looking little species, but apparently distinct from everything yet described. The anterior region of the elytra is more strongly than usual (in the genus) declivous towards the prothorax, and the lateral prominence of the shoulders (suggestive of some forms of *Leptops*),—slight but evident in this species,— is exceptional in *Oxyops*. This latter character seems to be unusually prevalent in the species that occur in the Northern Territory.

N. Territory of S. Australia; taken by Mr. J. P. Tepper.

O. MACULATA, sp.nov.

Sat lata; supra ferruginea, squamis fasciculatis in tuberculis nonnullis maculatim ornata; subtus picea. [Long. $3\frac{3}{5}$, lat. $1\frac{3}{5}$ lines.

The rostrum is somewhat gibbous near the apex. The head is deeply furrowed between the eyes, the space between the furrow and either eye being clothed with erect long white scales. The prothorax is coarsely rugulose (the base and front of the disc less coarsely than the other parts). The elytra are profoundly foveolate in close rows (the interstices granulate); the shoulders are protuberant laterally in such fashion that viewed from above a slight conical process appears to project beyond the lateral margin on either side; each elytron bears several tubercles which are topped with a conspicuous fascicle of erect white scales; the tubercles are arranged as follows,—on the third interstice an elongate one at the base, a small one before the middle, and a large one just above the posterior declivity,—on the 5th interstice a small one level with the middle one of the 3rd interstice, and another small one near the apex,—on the 9th interstice several small ones. The whole upper and under surface and the legs are thinly clothed with small adpressed white scales. The mesosternal projection is very well-defined and pointed.

Apparently near *O. niveospersa*, Pasc., (a species I am not acquainted with except by description) but differing in the shape of the rostrum, in the conspicuous crest of white scales on either side between the eyes, &c., &c.   The 2nd joint of the funiculus is nearly as long as the 1st and 3rd (which are equal each to the other) together.

Fraser Island ; sent by F. M. Bailey, Esq., of Brisbane.

MEDICASTA.

M. OBSCURA, sp.nov.

Fusca, griseo-squamulata, squamis in elytris fascias tres obscuras (1^{am} basalem, 2^{am} medianam, 3^{am} subapicalem) formantibus ; rostro in medio sulcato, basi sub-bilobato ; prothorace quam longiori vix latiori, antice angustato, ruguloso, lateribus a basi antrorsum ad medium subparallelis, a medio arcuatim angustato : elytris prothorace dimidio latioribus, subparallelis, striatis, striis profunde nec crebre punctulatis.   [Long. 2⅔, lat. 1 line.

I think this insect may be referred to *Medicasta*, though it presents some slight structural differences from the species on which the genus was founded : its general appearance, however, is very similar.

The rostrum is a little longer than the head, its basal portion longitudinally sulcate, the sides of the sulcation convex, clothed with pale setiform scales, and ending somewhat abruptly on the head nearly as far back as the level of the hind margin of the eyes. The antennæ are inserted at a distance from the front of the rostrum about equal to a third of its length : their scape is less than half as long as the funiculus, and reaches back to about the middle of the eye : the joints are proportioned much as in the description of *Medicasta :* the scrobes are as stated in the description of that genus, but hardly extend forward so far as I should expect.   The eyes are narrowed at their lower end, but can scarcely be called "acuminati."   The underside is clothed rather evenly but not closely with pale setiform scales.   The prothorax is densely clothed with rather pale scales, under which its surface appears to be confusedly rugulose.   The fasciæ into which the scales on the elytra are collected are not very conspicuous.

The present insect differs from *M. leucura*, Pasc., *inter alia* by the absence of tubercles on the elytra.

Northern Territory of S. Australia ; a single specimen taken by Mr. J. P. Tepper.

BELUS.

B. INSIPIDUS, sp.nov.

Niger, squamulis albidis variegatus ; prothorace canaliculato (canali albido-pubescenti), fortiter granulato (fere tuberculato) ; elytris crasse profunde subrugulose punctulatis, juxta suturam subdepressis, apice productis attenuatis, punctis parvis albido-hirtis confuse ornatis ; subtus sternis et latera versus segmentis ventralibus albido-hirsutis ; femoribus anticis obscure dentatis.

[Long. (rostr. incl.) 6⅔, lat. 1⅖ lines.

Very similar in shape to *B. hemistictus*, Germ., but with the antennæ very much shorter (they scarcely exceed the rostrum in length), the elytra slightly dilated immediately behind the middle, devoid of a carina, much more coarsely sculptured and gently convex longitudinally on either side of the suture, and much more confusedly sprinkled with spots (which are all small) of pale pubescence (these spots being scarcely more concentrated in one part than in another) ; also the underside is marked differently from that of *hemistictus*.

The rostrum is stout, cylindric, arched, shining, and finely punctulate throughout, being rather longer than the prothorax ; the head is very coarsely rugulose-punctulate, the orbits lined with pale pubescence ; the prothorax is at its base a little wider than its length down the middle, bears a wide well-defined longitudinal channel which is clothed with pale pubescence, and is sculptured even more coarsely than the head, the intervals between the punctures being quite tuberculiform ; the scutellum is clothed with pale pubescence. On the underside the median part of the sterna is thinly and the lateral thickly clothed with pale pubescence, the middle part of the ventral segments is glabrous and shining while a large spot of pale pubescence occupies either side of each segment, but these spots are scarcely united one with another into

the form of a vitta. The anterior femora are scarcely distinctly
dentate beneath. The pubescence on the specimen before me, which
is probably a female, is very pale brown rather than white, but
the specimen is not fresh.

N. Territory of S. Australia ; taken by Mr. J. P. Tepper.

LONGICORNES.

PACHYDISSUS (PLOCÆDERUS) AUSTRALASIÆ, Hope.

The collection made by Mr. J. P. Tepper, near Port Darwin,
includes a specimen (♂) of a *Cerambycid* which seems to agree
very well with Mr. Hope's description of this insect, except that
it is considerably larger (14½ lines) than the size there mentioned,
Considering the tendency of the *Cerambycidæ* to vary in size, I
think a difference of four lines in length not incompatible with
identity. The resemblance of this specimen to *P. sericus*, Newm.,
is excessively close, except in respect of the antennæ, which are
very different, being nearly twice the length of the body, and
having their joints differently proportioned ; the 3rd joint is
nearly half again as long as the 1st, the 4th equal to the 1st, the
3rd and 4th strongly (but not so strongly as in *P. sericus*) swollen
towards the apex, the 5th same length as 3rd, the 6th and remaining joints each longer and more slender than the joint next
before it.

PHORACANTHA.

P. FALLAX, Pasc.

The size of this species is given by its author as "10 lines." I
have a single specimen of that size but the average size is 8
lines.

TRYPHOCHARIA.

The genus *Tryphocharia* bears a considerable resemblance to
Phoracantha, from which Mr. Pascoe, its author, distinguishes it
by the small size of its prothorax in proportion to the elytra, by
its more linear femora, its forehead more narrowed in front, its

shorter antennæ, and especially by the spinose joints of the latter bearing *two* spines instead of one only. It may be added that the antennæ have a more or less distinct indication of a twelfth joint.

The genus *Xypeta* (formed by Mr. Pascoe at the same time as *Tryphocharia*, for an insect previously described by him as *Phoracantha grallaria*) appears to differ from *Tryphocharia* only by its forehead wider in front, its longer antennæ, and its shorter anterior and longer posterior legs. There can be little doubt, I should say, that *Phoracantha gigas*, Hope, should be placed in this genus, for though Mr. Hope's *description* gives very little information about the structural characters, the accompanying figure represents it as having long antennæ with two spines on each spinous joint, and posterior femora slender and much longer than those of *Tryphocharia*.

It is very likely that among the species described as *Phoracantha* there may be others attributable to *Tryphocharia*, and possibly to *Xypeta*. The description of *P. acanthocera*, Hope, reads much like that of a *Tryphocharia*, but as it contains no mention of the length of the antennæ, nor of the number of spines on their spinose joints, nor any statement of the size of the insect, no positive conclusion is possible without a re-examination of the type. It is much to be wished that those who possess any of the original types of the Australian species insufficiently described by the earlier authors would publish a full and minute description of the same in the Transactions of some Australian Society.

The following species attributed to the genus *Tryphocharia* I have not seen, and am satisfied are quite distinct from anything known to me, viz., *T. Mitchelli*, Hope ; *T. superans*, Pasc. ; and *T. Mastersi*, Pasc. The first of these is said to be found in N.S. Wales and Queensland ; from the description and figure it would appear to be characterized especially by the very small spine on either side of the prothorax, the elytra distinctly bispinose at the apex, and the markings of the latter, which are of a pale yellow colour, with the base, the suture, the lateral margins, the apex, and a transverse fascia behind the middle, dark fuscous. It is probable that these markings are variable, but not, I think, to an extent that would bring any species known to me near it. *Phoracantha superans* (from Tasmania) was originally characterized by Mr. Pascoe as having the spinose joints of the antennæ "armed with *a* spine at the apex," but when that gentleman formed the genus *Tryphocharia*, he placed *superans* in it, from which it would appear that the original description was defective. The sides of the prothorax in this insect are said to bear a slender elongate straight spine, and the elytra to terminate in two long acute spines, and to be of a pale fulvous yellow colour, with the base and margins dark chestnut-brown. The description also states that the elytra gradually decrease in size and proximity as they approach the apex, but this character would appear so improbable that there is doubtless some error in the statement which I conjecture should be read as applying to the words "punctures on the elytra" accidentally omitted. *T. Mastersi* seems to resemble *Odewahni*, but to have the apices of its elytra bispinose, the puncturation of the same less close, and the prothorax tubercled (not spined) at the sides.

The following species of *Tryphocharia* are, I believe, correctly named in my own collection, and some other collections to which I have access.

T. HAMATA, Newm.

♂. (*longipennis*, Hope), said to occur in N. S. Wales, Victoria, and Tasmania. My own specimen is from Western Australia. It (*i.e.*, my Western Australian specimen which, if compared with the original, might possibly prove distinct, though it agrees very well with the description such as it is) is of a rather dark brown colour with an obscure blackish fascia considerably in front of the middle of the elytra,—and the front of the lateral margins, the hinder half of the suture, and a kind of vitta occupying the hinder half of the disc of the same,—obscurely darker than the general colour, the interstices of the punctures (especially in a longitudinal direction) obscurely yellowish. The prothorax and breast are a good deal clothed with rather long pale brown woolly pubescence, the elytra being thinly sprinkled with pale hairs. The prothorax is the same width (from the base of one spine to that of the other) as it is long down the middle, with its upper surface a good deal flattened, and its sculpture of the character usual in the genus, its lateral spines long, slender and curved towards the elytra. The antennæ reach a little beyond the elytra ; their joints from the 4th inclusive extremely flattened (but not carinate on the upper face), joints 3-8 bearing two equal spines (one on each side) at the apex,—all the spines directed hindward rather than outward, and all small, the pseudo-twelfth joint short but rather well defined ; the elytra are truncated (rather obliquely) at the apex, each end of the truncation bearing a long sharp spine. The hind tibiæ are a little curved.

T. ODEWAHNI, Pasc.

In his description of this species its author states that its elytra have the apex " rounded," but in a figure (given by him sub_sequently in the Journal of Ent., Vol. II.) the apices of the elytra are represented as straightly truncate. I have never seen a *Tryphocharia* having rounded elytral apices, but the species that is most plentiful in South Australia (I have specimens from the far west, from Adelaide, and from the Victorian border) has the apices almost straightly truncate with the inner end of the truncation produced in a short sharp spine. The elytra also have an obscure rather large blackish spot on the disc a little in front of the middle, which is represented in the figure of *T. Odewahni* but not mentioned in the description, and which appears to be highly characteristic of the species. In other respects this insect agrees with both description and figure of *T. Odewahni* and, I have no doubt, is that species. It differs from *T. hamata* structurally in having the lateral spine of the prothorax smaller and straight (or nearly so), and the external end of the truncate apex of the elytra not spined. The surface of the prothorax is much flattened. The antennæ scarcely differ from those of *T. hamata* except in being a little shorter, with the pseudo-twelfth

joint less developed. The apical part of the elytra (as in
T. hamata) is punctured not at all faintly, though very much less
coarsely than the front part.

The following species appear to be new :—

T. PRINCEPS, sp.nov.

Robusta; minus parallela; fusca, antennis palpis pedibus
elytrisque testaceis, his fasciis ternis (basali, antemediana, et
postmediana) fuscis instructis; supra sparsim sat longe albido-
pubescens; subtus meso- et meta-sternis et segmentorum ventralium
parte postica sat dense aureo-pubescentibus; elytris antice fortiter
rugulose, postice gradatim subtilius obsoletius, punctulatis, apice
singulatim oblique truncatis et bispinosis; prothorace leviter
transverso, valde ruguloso, tuberculis 4 et spatio mediano
lanceolato lævibus instructo, lateribus spina forti instructis;
femoribus linearibus.   [Long. 19, lat. 5 lines.

In the specimen before me (which is a female) the antennæ are
decidedly shorter than the whole body, and have their joints 3-9
spined on either side (each less strongly than that preceding it ,
the 2 spines on each joint equal to each other, and much stronger
than those of *P. hamata* and *Odewahni*, joints 3-11 carinate above,
and the apical part of joint 11 simulating a twelfth joint. The
spine on either side of the prothorax is strong, not bent, and very
sharp. On the elytra none of the fasciæ quite touch the lateral
margins, and only the basal one touches the suture ; this (*i.e.* the
basal fascia) extends from shoulder to shoulder, and reaches back
about an eighth part of the distance to the apex of the elytra (there
is a little infuscation not connected with the fascia along the
front part of the lateral margin) ; the antemedian fascia is quite
narrow—almost linear—and somewhat of the form *N* ; the postmedian fascia is of a lighter brown than the other two, and
resembles the antemedian one somewhat in shape, but with a
blurred and less defined outline. The elytra are about $2\frac{1}{6}$
times as long as together wide, and about four times as long as
the prothorax ; they are slightly at their widest behind the middle,
and their sides are scarcely perceptibly incurved behind the base.

Allied apparently to *T. Mitchelli*, Hope, the description of which deals with little but colour; that insect, however, is said to be twelve lines in length, and to have a "minute" spine on either side of the prothorax; the description of the markings on the elytra ("variegated with brown spots") is too vague for identification, but, judging by the figure, *T. Mitchelli* has the base suture, lateral margins, and apex infuscate with a single elytral fascia, postmedian, and of very different shape from the postmedian fascia in the present insect. From the other previously described species, its more robust, massive form will at once separate *T. princeps*.

N. Territory of S. Australia; taken by Mr. J. P. Tepper.

T. UNCINATA, sp.nov.

♂. Minus robusta: sat parallela; fusca, antennis palpis pedibus elytrisque testaceis, his maculis ternis (basali antemediana et postmediana) fuscis instructis; supra sparsim sat longe albido-pubescens: subtus meso- et meta-sternis et segmentorum ventralium parte postica sat dense aureo-pubescentibus; elytris antice fortiter rugulose, postice gradatim subtilius obsoletius, punctulatis, apice singulatim recte truncatis et bispinosis; prothorace haud transverso, valde ruguloso, tuberculis 4 et spatio mediano lanceolato lævibus instructo, lateribus spina magna acuta hamata instructis; femoribus linearibus.   [Long. 18, lat. 5 lines.

The fuscous spots on the elytra are,—a small one at the base on either side of the scutellum, a small one immediately in front of the middle near the suture on either side, and an elongate larger one touching the suture about half-way between the middle of the elytra and the apex. In the specimen before me the antennæ reach back very slightly beyond the elytra; their structure scarcely differs from that of the preceding except in joints 6-11 only being distinctly carinate, joint 9 scarcely spined, and the spines on joints 3-5 much more robust with the inner spine very much feebler than the outer. The general form is distinctly more parallel and less convex than in the preceding, the prothorax

93

(of course exclusive of the spines), is not at all wider than long down the middle (in *princeps* it is nearly ¼ again as wide as long), and the apical truncation of the elytra runs straight across. The large sharp hooked spine on either side of the prothorax distinguishes this species from all previously described except *hamata*, Newm., from which it differs widely in size, colour, &c., &c. The sides of the prothorax are almost perfectly parallel from the base to the anterior constriction, which is very strongly defined. The posterior tibiæ are straight. The outer spines on joints 3 and 4 of the elytra are very much larger than in any other *Tryphocharia* known to me.

Found near Adelaide; rare.

T. PUNCTIPENNIS, sp.nov.

♂. Sat robusta; fusco-brunnea, elytris testaceis, basi summa et sutura postice infuscatis; prothorace et sterno griseo-sublanuginosis; abdomine breviter pubescenti; prothorace quam longiori vix latiori, fortiter ruguloso, tuberculis 4 obscuris et spatio mediano lanceolato instructo, lateribus spina gracili elongata vix arcuata instructis; elytris punctis fuscis antice magnis rotundatis postice parvis impressis, apice singulatim suboblique truncatis fortiter bispinosis; antennis manifeste 12 articulatis.

[Long. 13, lat. 3½ lines.

The head, prothorax, legs and antennæ are of an almost uniform dark reddish-fuscous colour, the elytra wholly testaceous with the exception of the base and the hinder half of the suture which are narrowly infuscate, and the punctures which are dark brown. On the front half of the elytra the punctures are large, round and isolated on the disc becoming evidently smaller towards the suture and lateral margins, on the apical half the punctures are comparatively fine and close but not at all faintly impressed. The antennæ agree in all respects with the description given above of those of *T. hamata* except in having the 12th joint well developed and perfectly distinct from the rest; it is about ½ as long as the 11th joint.

Apart from colour resembles the male of *T. hamata* described above, but with the lateral spines of the prothorax very nearly straight, the punctures of the front part of the elytra different (much more separated from each other by defined intervals), the hind tibiæ very nearly straight, and (especially) the 12th joint of the antennæ as distinct as any of the other joints.

Fowler's Bay; taken by Prof. Tate.

N.B.—A ♀ *Tryphocharia* taken at the same time and place by Prof. Tate is evidently this species though differing from the male as follows : much larger (20 lines), antennæ decidedly shorter than the body with the 12th joint very little developed; prothorax evidently narrowed from base to apex and having shorter lateral spines; elytra less strongly bispinose at the apex, each (in addition to the marking described above) with a large elongate fuscous blotch on the disc a little before the middle (probably this is an individual rather than a sexual character), legs and antennæ testaceous-brown.

COPTOCERCUS.

C. NIGRITULUS, sp.nov.

Nigro-piceus, elytris singulis macula parva antemediana, fascia lata mediana et macula magna apicali instructis ; his fere ad apicem fortiter punctulatis, apice emarginato-truncato, truncatura externe fortiter spinosa ; prothorace tuberculis 4 nitidis instructis, disco nitide lanceolato-elevato, lateribus obtuse tuberculatis.

[Long. 5, lat. 2⅔ lines.

The whole insect (except the yellow marks on the elytra, and the palpi which are reddish-brown) is almost unicolorous, the legs and antennæ having only a very slight reddish tone. The puncturation of the elytra is coarse and close at the base, becomes even more so about the middle, and in the apical third becomes closer and less strong to near the apex, and even there it can hardly be called obsolete. The prothorax is not longer than wide. On the elytra there is a very small obscure yellow spot between the margin and the antemedian spot, the postmedian fascia narrows from the lateral margin to the suture but does not quite touch either, the apical yellow space extends backward to about the level of the commencement of the apical sixth of the suture. The antennæ are considerably longer than the body in the specimen before me, and have their joints 3-7 spined at the inner apex.

A rather short robust species as compared with others of the genus. This character, combined with the nearly black antennæ and legs, the peculiar elytral puncturation and sub-transverse prothorax will distinguish it, I think, from all its Australian allies.

N. Territory of S. Australia ; taken by Mr. J. P. Tepper.

APROSICTUS.

A. INTRICATUS, sp. nov.

Fusco-brunneus, plus minus griseo-tomentosus ; elytris antice testaceo-rufis postice testaceo-brunneis, maculis fasciisque nonnulis piceo-nigris notatis, postice fortiter bispinosis ; prothorace fortiter ruguloso, macula discoidali lævi, longtitudine latitudini æquali.   [Long. 12, lat. 2⅓ lines.

On the elytra the anterior third of the lateral margin is broadly blackish, and the hinder part of this blackish space runs out in a fascia-like manner to the suture ; at the middle there is a strongly angulated narrow black fascia, from immediately behind which a black line runs down the middle of the disc about half-way to the apex.   The apex of all the femora is black, as also the inner half of the upper face of each of the 4 posterior femora. The whitish hairs are most dense on the prothorax, the apical third of the elytra, and the whole undersurface—on these parts being moderately close, on the rest of the surface very sparse. The interstices of the rugose sculpture of the prothorax are very nitid ; on the disc immediately behind the middle is an ill-defined, rounded space on which the rugosity and pilosity both fail, and which consequently appears as a shining spot.   There is some indication of a similar spot (as though several interstices coalesced) close to the base on either side of the middle.

This species appears to differ from the Malayan *A. Duivenbodii*, Kaup, *inter alia* in its bispinose elytral apices, in the absence of a glabrous longtitudinal line on the prothorax, and in the much more intricately patterned elytra, the apices of which are their most pale-coloured portion. It is an extremely interesting addition to the Australian fauna.

N. Territory of S. Australia ; a single male, taken by Mr. J. P. Tepper.

SCOLECOBROTUS.

S. SIMPLEX, sp.nov.

Elongatus, breviter sat dense pubescens ; brunneo-testaceus, femoribus posticis 4 apicem versus infuscatis ; prothorace quam latiori paullo longiori, antice angustato, transversim æqualiter rugato, ad latera pone medium subtuberculato, disco in medio utrinque minute tuberculato ; elytris antice crebre minus fortiter, postice subtilissime obsolete, punctulatis, apice rotundatis.

[Long. 8, lat. 1⅔ lines.

The head and prothorax of the example before me are a little darker and more reddish than the rest of the surface.   This species is closely allied to *S. Westwoodi*, Hope, from which it differs in its smaller size, testaceous *brown* colour, more sparse pubescence, much finer and and closer basal puncturation of the elytra which does not extend so far backward, rounded apices of elytra and infuscate hinder four femora.

I am doubtful of the sex of my example ; its antennæ are a little longer than the body, with a well-defined twelfth joint, the joints proportioned *inter se* much as in *S. Westwoodi;* they are feebly serrate owing to the apex of each being a little produced within, but the joints have not their inner edge cut into sharp teeth as in *S. Westwoodi*.   In all probability the specimen is a female.

N. Territory of S. Australia.

N.B.—A specimen taken by Mr. J. P. Tepper in the Northern Territory may be the male of the above, but it presents slight differences which suggest its representing a distinct closely allied species. Its prothorax is of a bright *reddish* testaceous colour and is somewhat more coarsely wrinkled transversely, with the dorsal tubercles scarcely traceable, and its elytra are rotundate-truncate at the apex rather than rounded.   It is, moreover, a little larger. Its antennæ are similar in length and in the proportion *inter se* of the joints, but they are altogether stouter and much more strongly serrated, though in the same manner as in my specimen and without any trace of the close serration that runs along the edge of each joint in *S. Westwoodi.*

S. VARIEGATUS, sp.nov.

Elongatus, breviter sat dense pubescens ; fusco-brunneus ; capite, prothorace, antennis, palpis, pedibusque brunneo rufis; elytris nigro adumbratis ; prothorace quam latiori sat longiori, antice angustato, transversim æqualiter rugato, ad latera pone medium obtuse tuberculato, disco in medio utrinque minute tuberculato ; elytris antice profunde rugulose, postice vix evidenter, punctulatis, apice fortiter bispinosis. Maris antennis corpore paullo longioribus (feminæ corpore paullo brevioribus), ut *S. Westwoodi* conformatis.

[Long. 10¾, lat. 2 lines.

The hinder four-fifths of the elytra are clouded with blackish immediately within the lateral margin.   The front part of this dark vitta is much the deepest in colour and is dilated so as nearly to reach the suture (in some examples more nearly than others) extending nearly (or quite) over the hinder half of the rugosely punctured space.

Port Lincoln, S. Australia ; on flowers of *Eucalyptus.*

ANTEROS.

This genus (characterized in 1845 by M. Blanchard on an undescribed Australian species) is probably identical with *Agapete* (characterized by Mr. Newman in the same year).   The diagnosis agrees very well with specimens of *Agapete* before me,—mentioning the very peculiar shape of the elytra and other characters.

At first sight it would appear as if the phrase " tarses à 1er article très court" were inconsistent with this supposition,—since the basal joint of the tarsi in *Agapete* is decidedly longer than the 2nd ; but the force of this objection disappears when it is borne in mind that the basal joint is decidedly shorter than the following two together, and that the genera with which M. Blanchard associates *Anteros* have the basal joint at least equal to the following two (that immediately after which M. Blanchard places it,—*Callisphyris*,—has that joint much longer than the 2nd and 3rd together). Thus compared the basal joint in *Agapete* would naturally be called " very short."

PARMENOMORPHA, gen.nov.

Gen. *Parmenæ* affinis, sed oculis crasse granulatis.

The description of *Parmena* in Lacordaire's Gen. des Col. IX. p. 275, exactly fits the small insect for which I propose this new name, with the single exception that the eyes (instead of being " subfinely ") are extremely strongly and coarsely facetted. The presence of a small, well-defined, triangular scutellum, and of a small sharp spine on either side of the prothorax, together with the smaller size of the basal ventral segment (very distinctly shorter than the following two together), will separate it from *Correstetha*, the strong sinus of the intermediate tibiæ from the Malayan *Dasyerrus*, the prothoracic spines from *Bybe*.

P. IRREGULARIS, sp.nov.

Testaceo-ferruginea, capite prothoraceque obscurioribus ; dense breviter pubescens et capillis longis erectis sparsim vestita ; antennis (♂. ?) corpore longioribus, sat robustis; capite prothoraceque rugulosis nec dense nec crasse punctulatis ; hoc utrinque pone medium spina laterali parva gracili instructo ; elytris lateribus basique fortiter nec crebre, disco crassissime sparsim, punctulatis.   [Long. 3, lat. 1 line.

The sculpture of the elytra is not unusual in the *Dorcadionidæ* —the inner middle part of the disc bearing a few very coarse punctures, while the remaining space is considerably more closely and less coarsely punctulate.

N. Territory of S. Australia ; taken by Mr. J. P. Tepper.

MICROTRAGUS.

M. JUNCTUS, sp.nov.

Angustus ; cinereo-variegatus, squamis nigrescentibus capil-lisque nigris intermixtis: prothorace rugoso ; elytris 4-costatis, costis externis apicem juxta, internis pone elytrorum medium, connectis.   [Long. 6, lat. 2 lines.

Head strongly convex ; prothorax not wider than down the middle long, its base and apex equal (the former bisinuate with the middle rather strongly angulated), its sides somewhat rounded and furnished behind the middle with a strong sharp projection the apex of which is scarcely bent hindward, its surface very convex and coarsely but not closely rugulose ; elytra with their humeral spines strong, sharp and bent, the four costæ (i.e., two on each elytron moderately strong and serrate rather than tuberculate, the inner pair meeting on the suture about two-thirds of its length from the base, the external pair meeting on the suture close to the apex, the space between the inner pair much flattened, the whole surface of the insect covered with rough dirty-looking brown scales mingled (especially along the costæ) with blackish scales and thinly sprinkled with rather long erect black hairs.

McDonnell Ranges, Central Australia ; taken by Mr. A. W. S. Wild.

LYCHROSIS.

M. Lacordaire [Gen. Col. IX. (2) p. 541] questions the generic identity of the two insects (one from Australia, the other from Sylhet), which Mr. Pascoe associated in this genus, and proceeds to furnish a diagnosis somewhat fuller than Mr. Pascoe's.   The Australian *L. luctuosus* does not altogether fit that diagnosis,—especially I do not find that the scape of the antennæ is of the peculiar form M. Lacordaire describes,—and it is very likely that the learned French author is right in thinking that two generic names are required. In that case the new name will have to take the place of *Lychrosis*, Lacord., as Mr. Pascoe founded his genus on the Australian species, for which, therefore, the original name must be retained.

I may add that I have before me several specimens of *L. luctuosus*, Pasc., taken by Mr. J. P. Tepper, near Port Darwin, which vary considerably in size ($4\frac{3}{5}$-6 lines), and also in markings, some of the white spots on the elytra showing much tendency to run together into connected lines.

HATHLIODES.

H. GRAMMICUS, Pasc.

Mr. Tepper's collection of *Coleoptera* from the N. Territory contains examples of a very variable species that appears to be this insect. The grey lines running down the elytra mentioned in the description of the type are seldom very distinct, and some-times quite untraceable, the whole surface being then evenly clothed with whitish pubescence. Abraded specimens (and judging by their frequency the pubescence seems to be very deciduous) are of an uniform shining ferruginous colour. In very fresh specimens the antennæ are evenly clothed with fine whitish pubescence, and their darker colour near the apex (mentioned in the description) is not noticeable. The length varies from $5\frac{1}{4}$ lines to 8 lines. Several of the specimens before me have traces of oblique striæ running between feeble rounded carinæ down the elytra (scarcely evident except in the apical half), and they may possibly represent a distinct species,* but I can find no other character to distingish them. The abruptly (*i.e.*, suddenly) narrowed apex of the elytra, not drawn out to a long point as in *H. lineella*, nor sub-emarginate as in *H. 4-lineata*, but separately obtusely pointed (in some examples separately rounded off with

---

* Possibly *H. moratus*, Pasc. The sharpness of the apex of the elytra seems to vary both in the striated and non-striated specimens.

scarcely a point), with the extreme apical margin thickened, seems to distinguish this species from all its North Australia congeners—unless *H. murinus*, Pasc., in the description of which the elytral apices are not characterized, and which is not known to me.

H. LACTEOLA, Hope.

In the above-mentioned collection there are also specimens of an insect which agree so well with the description of *H. lacteola*, Hope, that I can hardly doubt their identity with it. They belong, however, to *Mycerinopsis*, having antennæ considerably longer than the body in the male, and the intermediate tibiæ formed as in the *Apomecynides*. It must be near *M. uniformis*, Pascoe, from which, however, the elongate strongly narrowed apex of its elytra would seem to distinguish it. I may say that the specimens before me are all somewhat more *yellowish* in colour than Hope's description would lead one to expect, but they are all more or less abraded, and there are unabraded portions here and there quite decidedly of a milky white. Their size varies from 4 lines to 6 lines.

PHYTOPHAGA.

PSEUDOTOXOTUS, gen. nov.

Palporum maxillarium articulus ultimus oblongo-ovalis, apice obtusus.

Ligula membranacea, antice fortiter emarginata.

Oculi mediocres, rotundati, sat fortiter convexi, fortiter granulati.

Caput minus elongatum, postice manifeste angustatum.

Antennæ corpori longitudine æquales (♂.?) vel vix æquales (♀.?), ante oculos positæ, articulo ultimo appendiculato.

Prothoracis latitudo maxima juxta basin posita.

Coxæ anticæ anguste separatæ, intermediæ subcontiguæ.

Femora postica vix incrassata, apicem versus fortiter angustata, parte angustata acute dentata.

Corpus totum dense pubescens.

Differs *inter alia* from *Megamerus* in the shape of the apical joint of the maxillary palpi, from *Cheiloxena* in the nondentate sides of the prothorax, from *Duboulaia* in the strongly convex eyes, from *Prionesthis* in the dentate hind femora, from *Carpophagus, Diphanops, Mecynodera,* and *Ametalla* in the long antennæ, and from *Polyoptilus* in the dense clothing of pubescence.

P. LINEATA, sp.nov.

Sat elongata ; ferruginea ; dense albido-pubescens ; elytris costis 3 vel 4 latis obscuris instructis; his nonnihil denudatis, postice obsoletis.   [Long. 4-6, lat. 1¼-1¾ lines.

The structure of the head and its organs is almost exactly as in *Polyoptilus Lacordairei,* Germ. The surface is entirely clothed with dense whitish hair beneath which it appears to be finely punctulate. The basal joint of the antennæ is about equal to the third,—joint 2 short, 3 twice 2, 4 nearly twice 3, 5 scarcely longer than 4, 6 equal to 5, 7-11 successively longer, the appendiculate part of 11 very short. The prothorax closely resembles that of *Polyoptilus* in structure, the suture between the pronotum and prosternum running (as in that genus) on the underside but appearing more conspicuous ; the prothorax is as long as wide, its greatest width immediately in front of its base, its sides concave in the middle, and convergent in the extreme front, so that a little behind the front the segment is not much narrower than at its widest ; there is a denuded and slightly elevated narrow line (abbreviated at both ends) running down the middle ; the angles are all obsolete. The elytra are not at all punctulate-striate but (as far as I can observe under the dense pubescence) are rather closely punctured with a confused mixture of large and small punctures ; three or four ill-defined wide rounded costæ originate at or near the base but do not extend much beyond the middle of the elytra hindwards (very similar costæ exist in *Polyoptilus Lacordairei*), which are almost devoid of pubescence and thus show a ferruginous colour in contrast with the nearly white pubescence, giving the elytra the general appearance of being nearly white with several obscure wide reddish longitudinal vittæ in the anterior

two thirds of their length.   The legs are extremely like those of
Polyoptilus, but are a little longer (especially the tarsi) and more
slender.   The prosternal process, though very narrow (like a
knife-edge) distinctly separates the coxæ and bends down hindward
(visibly from behind) ; the mesosternal process on the other hand
can scarcely be traced distinctly between the intermediate coxæ,—
thus reversing the structure of *Polyoptilus* where the intermediate
coxæ are more distinctly separated than the anterior.   The basal
ventral segment is rather more than twice the length of the next
two together.

The resemblance of this insect to a *Toxotus* is most extraordinary.
N. Territory of S. Australia ; taken by Mr. J. P. Tepper.

DITROPIDUS.

D. PALMERSTONI, sp.nov.

Late ovatus ; æneus ; labro, antennarumque articulis primis sex
fulvis: articulo 1° robusto, 2° subgloboso, 3° elongato, 4°-6° sat
brevibus: capite prothoraceque crebre fortiter punctulatis ; elytris
punctulato-striatis, interstitiis planis (externis vix convexis) crebre
minus subtiliter punctulatis.   [Long. 1-1½, lat. ⅓-1½ lines.

A very wide almost semicircular species ; the even, close and
very strong puncturation of the head and prothorax, together with
the rather close and strong confused puncturation of the elytral
interstices, without any transverse strigosity, will distinguish it
from all others bearing a general resemblance to it.   Probably
D. laminatus, Chap., is its nearest ally from which it differs *inter
alia* in the clypeus not being bidentate (at least not in the specimen
before me), in the prothoracic puncturation being by no means
" aciculate," and in the even punctulate striation of the elytra.

N. Territory of S. Australia ; taken by Mr. J. P. Tepper.

IDIOCEPHALA.

The following species I believe to be *Aporocera catoxantha,*
described by Mr. Saunders on specimens from Port Essington.

Herr Suffrian has already pointed out that the species in question is probably inseparable from *Idiocephala*. I have a good many specimens before me, of which one only agrees with the description exactly in respect of colour and shape of markings.

I. CATOXANTHA, Saund., var. (?)

Oblongo-quadrata; flava vel ferruginea; antennis (maris corpore vix longioribus feminæ brevioribus), tibiis apice et tarsis picescentibus; elytris (spatio communi γ simulante et marginibus lateralibus ipsis exceptis) cyaneis; capite prothoraceque crassissime nec crebre punctulatis; illo longitudinaliter plus minus conspicue canaliculato; elytris fortiter subseriatim punctulatis.   [Long. 2-3¾, lat. 1-1¾ lines.

The γ-like mark on the elytra is very coarse and thick (as though daubed on with a coarse brush), the extremities of its arms nearly reaching the humeral callus on either side, and its foot being at the apex of the suture.

N. Territory of S. Australia; taken by several collectors.

I. PURA, sp. nov.

♂. Breviter oblongo-quadrata; flavo-rufa; antennarum articulis 6 ultimis et prothoracis margine basali summo nigricantibus: scutello elytrisque læte cyaneis: capite prothoraceque fortiter nec crebre, elytris crebre fortiter vix seriatim, punctulatis: tarsorum apice subinfuscato.   [Long. 1¾-1¾, lat. ¾-1 line.

N. Territory of S. Australia; collected by Mr. J. P. Tepper.

I. PALMERSTONI, sp. nov.

♂. Breviter oblongo-quadrata: rufa; antennis apicem versus, metasterno, abdomine, et scutello, nigris; elytris cyaneo-nigro variegatis capite leviter obscure, prothorace sparsim nec fortiter, elytris sat fortiter sat crebre subrugulose vix seriatim, punctulatis.   [Long. 1⅜, lat. ⅞ line.

The dark markings on the elytra are as follows: a blotch shaped like a subequilateral triangle with the front margin of the elytra as its base, and its apex on the suture a little behind the middle, a narrow edging to the hind part of the suture, and (on either side) a blotch of similar shape and size to that already mentioned, having as its base the hinder two-thirds of the lateral margin, and its apex falling on the suture a little behind the middle. Thus if the dark colouring be regarded as the ground tint of the elytra, there would appear to be on each elytron a broad subparallel red stripe running from the lateral margin (immediately behind the base) obliquely almost to the suture, and a red spot on the inner apical extremity not quite touching the suture.

The undersurface is thinly clothed with short silvery hairs.

N. Territory of S. Australia ; taken by Mr. J. P. Tepper.

APOROCERA.

The following species agrees sufficiently well with the description of *A. apicalis*, Saund., (from N. S. Wales), to prevent my giving it a new name. It appears to differ chiefly in the colour of the ventral segments (which in some examples is almost wholly red), in the elytra being narrowly margined in front with black, and in the scutellum not being margined with black.

A. APICALIS, Saund., var. (?)

Elongato-quadrata ; rufa ; antennis (late compressis, corpore brevioribus), elytrorum basi anguste et apice late, pygidio apice, metasterno, abdomine (vel toto vel in parte), femorum et tibiarum apice, tarsisque, nigricantibus ; capite prothoraceque crassissime acervatim, elytris profunde seriatim sat crebre, punctulatis.

[Long. 3, lat. 1⅔ lines.

The punctures on the head and prothorax are extremely large and deep ; they are placed on the anterior part of the former, and on the latter are almost confined to the oblique depressions usual in this genus, which run from near the front of the lateral margins to near the middle of the base. The dark apical cloud on the elytra occupies nearly the posterior quarter of those organs. The second joint of the antennæ is of a paler colour than the rest.

N. Territory of S. Australia ; collected by Mr. J. P. Tepper.

TERILLUS.

T. MICANS, sp.nov.

Oblongus ; convexus ; obscure fuscus vel piceo-ferrugineus, æneo- vel viridi-micans ; antennis ferrugineis ; capite, prothorace, et corpore subtus, pilis brevibus vestitis ; capite obscurius, prothorace crebre fortiter rugulose, elytris profunde crebre nec rugulose nec seriatim, punctulatis.   [Long. 3-3⅔, lat. 1⅓-1¼ lines.

The general colour is a kind of pitchy ferruginous much shot with pale greenish iridescence on the underside, the head and the prothorax.   The general colour of the elytra is of a more decidedly ferruginous tone than that of other parts of the body and their iridescence is coppery rather than green ; the antennæ are entirely ferruginous ; the legs vary from dark ferruginous to dark piceous in colour, the femora in many examples being æneous, and the tarsi rarely as darkly coloured as the tibiæ.   The puncturation of the head is close, rather fine, and very rugose, but much obscured by a clothing of short adpressed shining grey hairs.   The separation of the clypeus from the front is hardly traceable.   The basal joint of the antennæ is moderately stout, the 2nd much more slender and a little more than half as long, the 3rd more slender still and about equal to the 1st in length, the 4th and following joints scarcely longer, the apical four a little compressed and dilated.   The prothorax is a little less than half again as wide as long, the base a little less than half again as wide as the front margin, the sides rounded (not at all angulated) with their edges appearing crenulated owing to the rugosity of the puncturation of the surface, the hind angles acute, the front angles little marked. The scutellum is finely and not closely punctulate.   The surface of the elytra is quite free from rugosities, and shows scarcely any indication of transverse sculpture in any light, its puncturation

being close, deep, and well-defined.   The
tarsi is a little longer than the second.   T
externally, the channel (of the hind tibi
deep and wide at the apex.

N. Territory of S. Australia ; taken by

T. POLITUS, sp.nov

Oblongus; convexus; fuscus, viridi-læte
pedibusque testaceo - ferrugineis ; capite
subtus pilis brevibus vestitis ; capite cre
profunde crebrius nec rugulose (hoc qua
parte latiori), elytris profunde crebrius ne
punctulatis ; his sat manifeste transvers
apicem versus vix convexis.   [Lo:

A very pretty species ; on the head and
fuscous ground colour is almost lost in th
cence, which, on the elytra, is almost c
surface of the large fovea-like punctures, t
it is somewhat diffused over the base and e
the green iridescence is strongest on the p
gradually less noticeable hindward ; in s
sternum might almost be called " metall
qualification.

From *T. micans* the non-rugulose puncti
will distinguish this insect ; from *T. poros*
to resemble rather closely) it differs *inter*
being much less than twice as wide as long
considerably in size.

N. Territory of S. Australia ; taken b

minus perspicue transversim rugatis, longitudinaliter (parte tei
antica excepta) carinatis.　　　　　　　[Long. 2⅘, lat. 1⅖ lir

In this species the green gloss that more or less pervades
whole surface does not anywhere overpower the testaceous und
lying tint, though it is variable in respect of its intensity ɛ
distribution, being usually most conspicuous about the lateral a
sutural margins of the elytra. In size and shape it reseml
T. politus, but differs (apart from colour), in the prothorax hav
a fairly defined dorsal channel, in the puncturation of the sa
being strongly rugulose and in the sculpture of the elytra, wh
in front is a little more inclined to run in rows and a little m
inclined to rugulosity, but in the hinder two-thirds falls int(
perfectly longitudinal arrangement with well-defined and stron;
convex interstices; from *T. porosus* the small size, and prothoi
much less than twice as wide as long, will distinguish this inse

N. Territory of S. Australia; taken by Mr. J. P. Tepper.

Colaspis.

C. Palmerstoni, sp.nov.

Oblonga ; sat convexa ; subtus picea ; supra purpurea, cupı
micans ; labro, antennis basi, tibiis, tarsisque (nonnullis exem]
pedibus totis) rufescentibus ; supra sat fortiter, sat æqualiṭ
crebre subrugulose punctulata.　　　　　[Long. 2⅘, lat. 1 li

The separation between the clypeus and the front is scarc
marked, the latter having an obscure transverse impression
some examples *very* obscure) near its anterior margin. The ſ
thorax is slightly more than a third again as wide as long,
base not very much wider than the front margin ; the latɛ

angulated in some examples (in some not symmetrically on the two sides), in other examples their lateral curve is scarcely sinuate.

I see no reason to regard this insect as other than a true *Colaspis*. The anterior margin of its prothoracic episterna is not convex, the claws are appendiculate, the tibiæ not emarginate externally, the prosternum is truncate behind, the lateral borders of the prothorax (in some examples at least) are distinctly and subangularly undulous, the basal joint of the hind tarsi is equal to the following two together, the antennæ are slender and a little more than half the length of the body, with the apical joints only very slightly compressed. The hinder four tibiæ are channelled externally, the channel being deepened at the apex where the tibia is decidedly dilated, its external apical angle being well-defined and the apex itself obliquely truncate.

N. Territory of S. Australia ; taken by Mr. J. P. Tepper.

ACETINUS.

A. ÆQUALIS, sp.nov.

Ovalis; nitidus; æneus; antennis fuscis, basi testaceis; pedibus testaceis, tibiis tarsisque plus minus infuscatis ; capite prothoraceque subtilius, elytris sat fortiter, crebre punctulatis ; his vix manifeste quadri-costatis; interstitiis subtiliter sparsim punctulatis. [Long. 2, lat. 1₅ lines (vix).

The elytral punc'uration has scarcely any tendency to run in rows and is close and moderately strong, the interstices among the punctures having a distinct system of very fine and sparing puncturation ; the elytral costæ are scarcely raised above the surface and would probably be quite untraceable were not the puncturation more or less interrupted by them. In some examples the underside is of a decided green colour. The sides of the prothorax are nearly straight and show no trace of dentation or unevenness.

Judging from M. Boisduval's very succinct description of his *A. (Colaspis) Australis* the present species differs from it in not being of a copper-colour, and in the hind body not being ferruginous,

—but as the size of *Australis* is not stated and the only information given (besides the description of colour) is that the upper-surface is everywhere punctulate, it is likely enough that there are many other points of difference. *A. æqualis* is much smaller than the other previously described species of *Agetinus*; as compared with *A. corinthus* and *subcostatus*, moreover, the sculpture of its elytra is altogether finer and smoother. I have not seen *A. jugularis*, Er., but from the description that insect appears (apart from size) to differ from *A. æqualis, inter alia*, in having the underside of the head rufous and the sides of the elytra transversely rugose.

N. Territory of S. Australia; taken by Mr J. P. Tepper.

SCELODONTA.

S. SIMONI, Baly.

Among the specimens collected in the Northern Territory by Mr. J. P. Tepper is an example of this genus which appears to be too close to *S. Simoni* to be wisely described as new ; nevertheless it differs from the description of that species in having the elytra and the sides of the prothorax marked with some rather conspicuous golden spots, and it is probable that if it were placed side by side with Mr. Baly's insect, it would be found to differ in other respects. This spotted var. (if it be a var.) may perhaps not unsuitably be distinguished by a local name ; I shall therefore propose to call it var. ? *Palmerstoni*. It may be noted that in Mr. Masters' "Catalogue of Australian Coleoptera" the generic name *Scelodonta* is omitted, making S. *Simoni* appear as a *Tomyris*.

RHYPARIDA.

R. ÆNEO-TINCTA, sp.nov.

Elongato-ovata ; nitida ; rufa ; capite, prothorace antice, elytrorum regione suturali antice, meso- et meta-sternis, femoribusque æneo-viridi-micantibus ; antennis (basi excepta), tibiis tarsisque fuscis ; capite prothoraceque subtiliter coriaceis ; clypeo distincte minus crebre, vertice prothoraceque leviter sparsim, punctulatis ; scutello subtiliter coriaceo impunctulato ; elytris sat fortiter (postice levius) punctulato-striatis, interstitiis sparsim subtiliter punctulatis ; femoribus posticis inermibus.   [Long. 3⅔, lat. 1⅜ lines.

The æneous colouring on the prothorax is confined to the front where it is obscure and cloudy.   The separation between the clypeus and front is indicated only by the difference in puncturation.   There is a distinct longitudinal sulcus between the eyes which in front meets a very ill-defined curved transverse impression.   The prothorax is a little more than half again as wide as long, its sides are rather strongly rounded, and its apical margin is considerably narrower than the base.   The minutely coriaceous surface of the head and prothorax renders them sub-opaque; the elytra are very nitid.   The green colouring is at its brightest on the elytra, where it occupies the whole space between the fourth striæ on either side extending backward nearly half-way to the apex.

N. Territory of S. Australia : a single specimen taken by Mr. J. P. Tepper.

R. MEDIOPICTA, sp.nov.

Elongato - ovalis : nitida : rufa ; antennis basi, mandibulis, genubus, tibiis apice, et tarsis, piceis ; elytris æneo-nigris, margine laterali antice et sutura (antice plus minus anguste, postice subito late sed interrupte) rufis ; clypeo sat fortiter sat crebre, vertice leviter sparsim, prothorace vix manifeste, punctulatis, scutello impunctulato ; elytris fortiter (postice multo levius) punctulato-striatis : interstitiis vix manifeste punctulatis ; femoribus posticis inermibus.   [Long. 3¼, lat. 1⅜ lines.

var. E'ytris nigris vix subæneis, margine basali, et sutura margineque laterali antice, anguste rufis.

The rufous colouring on the elytra is extremely variable.   The extreme basal margin and the anterior two-fifths of the suture appear to be always rufous, but in some examples much more narrowly than in others ; at two-fifths of the length of the suture the rufous colouring spreads out into an elongate-oval spot, which nearly or quite reaches the apex, but at about three-fifths of its length the suture becomes dark again and appears as a narrow stripe dividing the hinder part of the rufous spot; the anterior three-fifths of the lateral margins also are rufous,—in some examples rather widely (especially towards the front), in others very narrowly. Of the antennæ the basal three joints (and in some specimens the base of the fourth) are rufous; the palpi are tipped with piceous. The clypeus is separated from the front by a transverse furrow, and the latter bears a longitudinal furrow. The prothorax is slightly more than half again as wide as long; its sides are nearly straight in their hinder two-thirds, then a little rounded and converging to the apex; the base is about a third again as wide as the front margin; the angles are all acute and pointed outward.

N. Territory of S. Australia; taken by Mr. J. P. Tepper.

R. AMPLICOLLIS, sp.nov.

Ovata; nitida; rufa; genubus nigris; elytris regione suturali antice late subæneis, latera versus nonnullis exemplis longitudinaliter æneo-notatis; clypeo fortiter sat crebre, fronte sparsim subtiliter, prothorace fortius nec sparsim (antice subtiliter, ad latera ipsa vix distincte) punctulatis; scutello fere impunctulato; elytris sat fortiter punctulato-striatis, striis internis antice leviter impressis, interstitiis subtilissime punctulatis; femoribus posticis inermibus. [Long. 3¾, lat. 2¼ lines (vix).

A very short wide insect. The clypeus is separated from the front by a well-defined furrow, the latter being longitudinally channelled. The prothorax is very nearly twice as wide as down the middle it is long; its sides are strongly rounded, its angles all acute and pointed outward; its base about a third wider than its front margin; the marginal portion of the surface all round (most widely at the sides) is nearly without punctures; on the punctured space thus enclosed the punctures are quite strong and close on the sides but become somewhat finer towards the middle. On the elytra the striæ are scarcely impressed except near the sides and apex, the punctures in the striæ being, however, much finer towards the apex than in front.

N. Territory of S. Australia ; taken by Mr. J. P. Tepper.

R. PUNCTULATA, sp.nov.

Sat late oblonga ; nitida ; fusco-rufa ; clypeo crebre fortius, fronte crebre subtilius, prothorace disco fortiter sat crebre latera versus etiam magis fortiter crebre, scutello obscure, punctulatis ; elytris antice punctulato-striatis, postice sublævibus ; femoribus posticis subtus leviter dentatis.   [Long. 3, lat. 1⅓ lines.

The clypeus is separated from the front by a transverse furrow and the latter bears a short longitudinal channel anteriorly.   The prothorax is rather small in proportion to the elytra ; its width is rather more than half again its length ; sides gently rounded ; all the angles acute, the anterior unusually produced in a forward and outward direction, the base not much more than a fifth wider than the front.   The striæ on the elytra are scarcely impressed but bear strong and rather close punctures, both striæ and punctures being nearly effaced in the hinder half ; the interstices are rather sparingly but very distinctly punctured.   The tooth on the underside of the hind femora is not much more than an angulation of the outline a little before the apex.

N. Territory of S. Australia ; taken by Mr. J. P. Tepper.

R. POSTICALIS, sp.nov.

Elongato-ovata ; minus nitida ; fulva ; capite inter oculos (non-nullis exemplis) elytrorum sutura (antice anguste, postice late dilatatim), meso- et meta-sternis (parte media excepta) et (non-nullis exemplis) femoribus plus minus evidenter, obscure viridibus ; antennis (basi excepta) piceis ; capite leviter sparsim, prothorace scutelloque vix manifeste, punctulatis ; elytris subtilius sat æqualiter punctulato-striatis, femoribus posticis inermibus.

[Long. 3¾, lat. 1⅓ lines.

The suture is narrowly greenish (in some examples scarcely so in the extreme front) to about the middle where the green colouring begins to dilate and forms an elongate-oval spot extending to the apex and reaching laterally to about the fifth stria on either side. The clypeus is continuous with the hinder part of the head, which bears an elongate notch-like impression between the eyes. The prothorax is about twice as wide as long, its base nearly twice as wide as its front margin, sides straight, front angles acute, hind scarcely so. The interstices of the striæ on the elytra are not punctured. The whole insect has a semiopaque appearance on the upper surface, and is minutely coriaceous.

N. Territory of S. Australia; taken by Mr. J. P. Tepper; also in my collection.

R. PICEITARSIS, sp.nov.

Sat late oblonga ; nitida ; fusco-testacea ; antennis basi excepta, genubus (obscure), tarsis, et abdominis apice fusco-nigris ; clypeo sparsim fortiter, fronte prothoraceque vix manifeste, punctulatis ; elytris sat subtiliter (postice etiam magis subtiliter) punctulato-striatis ; femoribus posticis inermibus. [Long. $2\frac{3}{5}$, lat. $1\frac{2}{5}$ lines.

The clypeus is continuous with the hinder part of the head which is finely canaliculated longitudinally. The prothorax is two-thirds wider than long down the middle, its base a little wider than the front margin, the sides rather strongly rounded, all the angles acute and pointed outward. The interstices of the striæ on the elytra are very finely and moderately thickly punctulate. The extent of dark colouring on the ventral segments is variable, being extended in some examples over more than the apical half.

N. Territory of S. Australia ; taken by Mr. J. P. Tepper.

R. UNIFORMIS, sp.nov.

Elongato-subovata ; nitida ; cyanea ; antennis basi et pedibus (nonnullis exemplis femoribus tarsisque piceis) rufis ; capite (clypeo paullo magis fortiter) prothoraceque sparsim subtiliter (hoc nonnullis exemplis vix perspicue) punctulatis ; scutello subtiliter coriaceo ; elytris (antice sat fortiter, postice gradatim subtilius) punctulato-striatis ; femoribus posticis inermibus.

[Long. 3½, lat. 1⅔ lines.

? hujus speciei var. Ænea, punctis in elytrorum striis majoribus et magis remotis.

The clypeus is continuous with the hinder part of the head, which bears a longitudinal furrow ; this furrow deepens and widens forward, being more or less forked at its apex, so that in some specimens there appears to be a pseudo-separation from the clypeus all the more apparent in occasional specimens with the clypeal puncturation a little stronger than usual. The prothorax is about half again as wide as long and is nearly twice as wide at the base as across the front margin ; the sides are scarcely rounded, the anterior angles acute and pointed outward, the posterior scarcely so. The interstices of the striæ on the clytra are sparingly and very finely punctured.

I do not feel sure that the "var. ?" mentioned above is not a very closely allied distinct species ; in addition to the differences already specified the sides of the prothorax are a little more rounded.

N. Territory of S. Australia ; taken by Mr. J. P. Tepper.

R. HERBACEA, sp.nov.

Late ovalis, postice sat angustata ; minus nitida ; supra subtiliter coriacea ; viridis ; antennis (basi rufa excepta), tarsisque, piceis ; labro, prothorace, scutello, pro- et meso-sternis, nonnullis exemplis metasterno latera versus, coxis, femoribus, tibiisque, rufis ; capite leviter, prothorace levissime, sparsim punctulatis ; scutello impunctulato ; elytris subtiliter (postice etiam magis subtiliter) punctulato-striatis, interstitiis haud punctulatis ; femoribus posticis inermibus.   [Long. 3, lat. 1¾ lines.

The clypeus is continuous with the hinder part of the head which bears a longitudinal fovea deepening and widening forward. The prothorax is nearly twice as wide as long, its base nearly twice as wide as its front margin, front angles acute, hind scarcely so, sides straight. The rather light green, silky appearance of the elytra gives this species a very distinctive appearance. In some examples the scutellum is more or less tinged with green.

N. Territory of S. Australia ; taken by Mr. J. P. Tepper.

R. SATELLES, sp.nov.

Late ovalis ; sat nitida ; rufa ; antennis (apicem versus) tarsisque piceis ; elytris, abdomine (apice plus minus rufescenti excepto) et nonnullis exemplis metasterno plus minus late, cyaneis vel viridibus ; clypeo sparsius subfortiter, fronte, prothorace scutelloque vix manifeste (sub lente forti sparsim subtilissime) punctulatis ; elytris distincte (postice gradatim subtilius) punctulato-striatis ; femoribus posticis inermibus.   [Long. 3, lat. 1⅔ lines.

The clypeus is continuous with the hinder part of the head, which bears a longitudinal channel more or less foveiform in front. The prothorax is nearly half again as wide as long, and its base is half again as wide as its front, all the angles acute and pointed outwards, sides gently rounded (most strongly in front). The head, prothorax and scutellum are finely coriaceous and sub-opaque, the interstices of the striæ on the elytra scarcely visibly punctured.

Resembles the preceding but is less narrowed behind, with the sides of the prothorax rounded, elytra nitid, &c.

N. Territory of S. Australia; taken by Mr. J. P. Tepper.

R. DISCOPUNCTULATA, sp.nov.

Lata ; ovata ; nitida ; cyaneo-nigra ; clypeo, labro, palpis, antennis, pedibusque plus minus rufescentibus ; clypeo fortiter rugulose, fronte crebre distincte, prothoracis disco fortiter crebre, punctulatis ; elytris sat fortiter punctulato-striatis ; femoribus posticis inermibus.   [Long. 2⅖, lat. 1¾ lines.

The clypeus is not truly separated from the front by a transverse furrow, but a longitudinal channel running down the latter spreads out in front in a manner that gives somewhat the appearance of a dividing furrow. The prothorax is nearly twice as wide as long, its base not half again as wide as its front margin, the front angles (though small) acute and pointed outward, the hind hardly so, the sides gently rounded ; the puncturation is very strong and rather close, and is mixed with a very different system of faint sparse punctures ; the coarse puncturation does not extend to the edges of the surface, being completely surrounded with a rather narrow strip on which there is only the shallow sparse puncturation scarcely visible save under a strong lens. The scutellum is impunctate. The striæ of the elytra are scarcely impressed in front, but their punctures are there large and deep ; hindward the striæ become more distinctly impressed and the punctures finer ; the interstices are rather closely and distinctly (though finely) punctulate.

N. Territory of S Australia ; taken by Mr. J. P. Tepper.

R. MOROSA, Jac.

I have before me several examples from the N. Territory which *may* be this insect. They agree fairly well with the description but are scarcely so *widely* ovate as I should expect, and are *greenish-* (not *bluish-*) black in colour. In all probability they represent a distinct species but if so it is one that it would not be wise to name without a comparison of specimens, and I therefore abstain from naming it.

R. INTERIORIS, sp.nov.

Ovata ; nitida ; testaceo-rufa ; clypeo sat fortiter rugulose, fronte sparsius obscure, prothorace sat fortiter subrugulose, punctulatis ; elytris (antice distincte, postice obsolete) punctulato-striatis ; femoribus posticis vix subdentatis.

[Long. 1¾, lat. ⅘ line.

In the example before me the head, prothorax, and all the underside (including the coxæ) are of a decidedly *reddish* tone, the antennæ, elytra, and legs being pale testaceous, but probably the shades of colour might vary in other specimens. The clypeus is continuous with the hinder part of the head, which bears a very fine longitudinal impressed line. The prothorax is about half again as wide as long, its base not much wider than the front margin; all the angles are acute, the sides rather strongly rounded. The elytral striæ are scarcely impressed in front, but are very distinctly set with rather small punctures ; towards the apex both striæ and punctures are subobsolete ; the interstices are quite devoid of puncturation.

The eyes are large and less separated than usual in the genus, the interval between them being less than the length of their shortest diameter. The apical five joints of the antennæ are more incrassated also than usual. The hind femora are not really dentate, but the attenuation of the apical portion is very sudden, so that the outline at this point is subangular. The hinder part of the head is a little tumid in appearance.

This and several other species in my collection appear to me very doubtfully congeneric with typical *Rhyparida*,—but they are at least closely connected with insects that have been attributed to the genus (*R. minuta*, Jac., e.g.), and present all the essential characters,—prothoracic episterna not convex,—posterior four tibiæ emarginate near external apex, and claws well developed and bifid.

I obtained a single specimen on Eucalyptus at Leigh Creek, about a hundred and fifty miles north of Port Augusta.

The following tabulation of the species of *Rhyparida* described above will perhaps be useful :—

A. Hind femora unarmed.

  B. Sides of the prothorax more or less rounded.

    C. Clypeus not separated from the front by a distinct furrow.

      D. Puncturation of prothorax not (or scarcely) defined.

        E. Elytra wholly testaceous; size small... *piceitarsis.*

        EE. Elytra wholly cyaneous or æneous.

          F. Prothorax and elytra unicolorous... *uniformis.*

          FF. Prothorax red..................... .. *satelles.*

EEE. Elytra testaceous with green
     markings .... *œneotincta.*

DD. Puncturation of prothorax very strong *discopunctata.*

CC. Clypeus separated from the front by a
     well-defined fovea.

    D. Antennæ (except at base) black or
     nearly so............................... *mediopicta.*

    DD. Antennæ wholly red or fuscous red.... *amplicollis.*

BB. Sides of the prothorax quite straight.

    C. Elytra entirely green .... *herbacea.*

    CC. Elytra fulvous with greenish marking..... *posticalis.*

AA. Hind femora toothed .... *punctulata.*

AAA. Hind femora scarcely toothed ; size under
     2 lines (i.e. much smaller than any of the
     preceding.......... .... *interioris.*

AUGOMELA.

A. ACERVATA, sp.nov.

Oblonga; convexa; pernitida; supra viridi-aurea, violaceo-varie-
gata; subtus violacea, viridi-aureo-variegata; elytris seriatim
punctulatis, seriebus medianis confusis; prothorace acervatim
punctulato.   [Long. 3½, lat. 2 lines.

On the upper surface the violet colour is spread over the back
of the head, the greater part of the thorax except the front and
sides, and a vitta-like space down the middle of each elytron
(commencing at a distance from the front equal to a fifth of the
whole length) which is strongly dilated in its front part; the
violet spaces are all edged with pure green.   The head is strongly
and rather closely punctured; the puncturation of the prothorax
is strong and rather close but condensed in patches, not however
more conspicuous on the sides than on the disc; on the elytra the
rows of punctures are more or less confused on the violet discal
space; the interstices are impunctate.   On the underside the

greenish golden colour is confined to the sides and middle of the prosternum and is not always present. The legs are of a deep violet colour, the antennæ blackish with their base pitchy testaceous.

N. Territory of S. Australia ; collected by Mr. J. P. Tepper and others.

N.B.—The above insect would seem to differ by its less rounded form from the hitherto described species of *Augomela*, which it approximates however by the style of its colouring and markings and by the form of its claws, as also of its prosternum ; its antennæ resemble those of *Calomela*. Possibly some authors might consider it the type of a new genus, but I think no great violence is required to associate it with *Augomela*.

CALOMELA.

C. APICALIS, sp.nov.

Lata ; oblonga ; convexa ; nitida ; rufa ; antennis (basi excepta) nigro-piceis ; elytris (margine laterali excepta) cyaneo-nigris, puncturis violaceis ; abdomine (segmento apicali excepto) cyaneo vel viridi ; elytris subseriatim, prothorace acervatim, punctulatis.

[Long. 3½, lat. 1⅜ lines.

The general colour of the elytra is black with a scarcely perceptible bluish tone, but the punctures, though fine, are evidently of a decided blue. The head is rather strongly (but not coarsely) punctured in front, nearly smooth behind. The prothorax is very coarsely punctured at the sides and has some connected clusters of finer punctures about the base and the middle of the disc. The elytra are finely punctulate, the punctures scarcely running in rows except near the apex (where they are very faint).

This species must resemble *C. cingulata*, Baly, from N.W. Australia, but differs *inter alia* in the elytra not being " *cyanea*," or in the least *striated*, and in the colour of the ventral segments which are entirely metallic green or blue save the front margin of the basal segment and the whole of the apical one which are bright red. I have five specimens before me all quite identical. Even if it be a local var. of *C. cingulata* it seems deserving of a name.

N. Territory of S. Australia; taken by Mr. J. P. Tepper.

C. PUNCTIPES, Germ.

This species is generally regarded as a form of *Curtisi*, Kirby, but I am unable to consider it so. I have before me a long series from widely separated parts of S. Australia which show very little variety *inter se* but invariably differ from typical *Curtisi* in having the prothorax wider and shorter with its disc much more coarsely punctured. Their differences *inter se* are almost confined to variations in the markings of the prothorax. I believe *C. punctipes* to be a good species.

C. DISTINGUENDA, sp.nov.

Oblonga ; convexa ; sat nitida ; rufa ; antennis (basi excepta) tarsisque piceis ; elytrorum vitta discoidali (antice abrupte dilatata), femoribus externe et tibiis cyaneis ; capite antice crasse postice subtiliter sparsim, prothorace ad latera crasse disco subtilius acervatim, punctulatis ; elytris subtilius punctulatis, puncturis latera et suturam versus seriatim dispositis, illic crassioribus.

[Long. 2⅖, lat. 1¾ lines.

Allied to *C. Curtisi*, Kirby. Compared with it the prothorax is not quite so short and is decidedly more thinly punctulate on the disc ; the elytra are much more finely punctulate and bear a differently shaped vitta, which is much narrower, and is abruptly dilated in front on its inner side ; and the underside, scutellum and thorax are entirely rufous, while the tibiae are wholly cyaneous.

N. Territory of S. Australia ; taken by Mr. J. P. Tepper.

C. TARSALIS, sp.nov.

Lata ; oblonga ; convexa ; sat nitida ; testacea vel rufo-testacea ; antennis (basi excepta) tibiis apice et tarsis nigris ; elytris regulariter seriatim, prothorace crasse (praesertim lateribus), capite crasse confuse, punctulatis.   [Long. 3, lat. 1⅘ lines (vix).

The puncturation of the prothorax is strong and by no means sparse on the disc, and becomes close and extremely coarse on the sides. The punctures in the rows of the elytra are rather large and strong and somewhat quadrate in shape ; the interstices are not convex, and are sparingly and very finely punctured.

Allied to *C. pallida*, Baly, and *geniculata*, Baly, both of which, however, are narrow insects with the disc of the prothorax finely punctured, the former having the legs entirely testaceous and the latter having black knees.

CHALCOMELA.

C. EXIMIA, Baly.

A few specimens agreeing very well with the description and figure of this insect were taken by Mr. J. P. Tepper near Palmerston (N. Terr.). Its precise habitat has not I think been known with certainty hitherto.

AMPHIMELA.

A. AUSTRALIS, sp.nov.

Late ovalis ; vix perspicue punctulata ; nitida ; nigra ; prothorace latera versus late testaceo ; antennis basi pedibusque plus minus picescentibus. [Long. 1¾, lat. ⅔ line (vix).

The antennae are scarcely so long as the head and prothorax together, joint 1 long and stout, 2 subglobular, 3 slender and nearly as long as 1, 4-6 short, 7-11 much wider and forming a cylindrical club. The antennae are inserted very far apart and close to the internal margin of the eyes. The head bears a longitudinal furrow on either side close within the eye, and an obscure median fovea. The eyes are large, rather coarsely granulated, and very convex. The prothorax is about three times as wide as long, very strongly convex transversely, narrower in front than behind, its anterior lateral portion consisting of a large tumid projecting lump which is cut off from the rest of the segment by a deep oblique sulcus ; the hind angles are obtuse, the base strongly lobed backward all across ; the testaceous margin on either side is wider than the black central portion ; under a powerful lens the surface is seen to be lightly and sparingly punctulate and to bear on either side near the margin an oblique furrow running forward from the base, the portion outside this furrow being tumid. The scutellum is minute and strongly transverse. The elytra are at their widest in front of the middle where they are a third again as wide as the prothorax, of which they are about four times the length ; they are rather attenuate towards the apex and are very strongly and sinuately contracted externally from a little behind the shoulder (apparently in order to accommodate the enormously developed hind femora). Their puncturation resembles that of the prothorax but with the addition here and there (especially towards the sides) of some rather stronger punctures. The anterior coxæ are strongly prominent, and almost contiguous, with their cavities closed behind.* The hind femora are as largely developed as in *Arsipoda*, and are unarmed ; the hind tibiæ are somewhat flexuous, and are strongly channelled and denticulate on their external margin, and mucronate at their apex ; their tarsi are inserted slightly above the apex (feebly after the manner of *Psylliodes*) and have the basal joint equal in length to the remaining three together ; the claws are appendiculate. The basal ventral segment is very strongly sulcate down the middle (this latter character probably sexual).

This remarkable little *Halticid* seems to be certainly very close to the East Indian *Amphimela* (of which I have never seen a type) though probably different enough to justify generic separation. Its agreement with *Amphimela* in the extraordinary position of its antennæ renders it convenient to refer it for the present to that genus which M. Chapuis (its author) regards as constituting a distinct "groupe" of the *Halticides*.

A single specimen sent by F. M. Bailey, Esq., and taken by him near Brisbane.

---

* I feel practically certain that this is the case, although I have not been able to dissect a specimen ; the example described is in a fairly satisfactory condition for examination.

Nisotra.

N. unicolor, sp.nov.

Ovata; nitida; testacea; antennis (basi excepta) piceis; capite impunctato; prothorace subtilissime, elytris sat fortiter, crebre punctulatis, his disco obscure subtiliter 3 vel 4 costatis, latera versus sat fortiter longitudinaliter sulcatis. [Long. 2, lat. 1½ lines.

The prothorax is quite twice and a half as wide as it is long down the middle; its sides are rather strongly rounded and sinuous immediately behind the prominent front angles, which gives them a slightly outward direction; there is a curved impression on either side near the lateral margin; the anterior and posterior longitudinal impressions are rather feeble. The pseudo-costæ on the elytra are little more than very fine lines appearing paler than the general colour, and interrupting the puncturation which is moderately strong and scarcely tending to a linear arrangement; the lateral sulcus on each elytron is strong, but does not extend much beyond the middle.

A very distinct species. The entirely different colour will at once separate it from its Australian congeners.

N. Territory of S. Australia; taken by Mr. J. P. Tepper.

Haltica.

H. Australis, sp.nov.

Supra nitida, cærulea; subtus cyaneo-nigra, breviter pubescens, antennis tibiis tarsisque fusco-piceis; capite inæquali, inter antennas longitudinaliter carinato, vix evidenter punctulato; prothorace quam longiori paullo latiori, pone medium transversim sat fortiter sulcato (sulco margines laterales attingente), disco vix evidenter ad latera sparsim subtiliter punctulato; scutello lævi; elytris crebrius subtilius punctulatis. [Long. 2¾-2¾, lat. 1¼ lines.

Extremely like the European *H. pusilla*, Duf., from which it differs as follows:—the antennæ are stouter and (with the tibiæ and tarsi) are of a more brownish colour; the prothorax is

95

longer in proportion to its width, and is a little more narrowed in front ; the eyes also are a little more prominent.

N. Territory of S. Australia ; taken by Mr. J. P. Tepper.

H. IGNEA, sp.nov.

Supra nitida, igneo-cuprea, prothorace obscure viridi-iridescente; subtus obscure æneo-picea, breviter pubescens, antennis pedibusque fusco-piceis ; capite inæquali, inter antennas longitudinaliter carinato, vix evidenter punctulato ; prothorace quam longiori parum latiori, pone medium transversim sat fortiter sulcato (sulco margines laterales attigente), disco vix evidenter ad latera sparsim subtiliter punctulato ; scutello lævi ; elytris crebrius subtilius punctulatis, latera versus sulco longitudinali sat fortiter impresso.

[Long. 3-3$\frac{1}{5}$, lat. 1$\frac{3}{4}$ lines (vix).

The elytral furrow is strong and conspicuous, commencing just behind the humeral callus and reaching to about the middle of the elytra. This furrow, together with the even longer prothorax and different colour, will distinguish this species from the preceding which in other respects it closely resembles.

N. Territory of S. Australia ; taken by Mr. J. P. Tepper.

H. FERRUGINIS, sp.nov.

Testaceo-ferruginea ; antennis (basi excepta) tarsisque piceis tibiis (anticis 4 leviter, posticis conspicue) et femoribus posticis apicem versus (nonnullis exemplis vix manifeste), infuscatis; capite inter oculos longitudinaliter postice canaliculato antice obscure carinato ; prothorace quam longiori dimidia parte latiori, basin versus transversim sat fortiter sulcato (sulco margines laterales attingente), vix perspicue punctulato ; scutello lævi ; elytris crebre subtilius (nonnullis exemplis subrugulose) punctulatis.

[Long. 3$\frac{1}{5}$, lat. 1$\frac{2}{3}$ lines.

N. Territory of South Australia ; taken by Mr. J. P. Tepper.

N.B.—The preceding three species all seem to agree perfectly with *Haltica* (*Graptodera*) and I fail to find any character on which to regard them as belonging to a distinct genus. As stated above, *H. Australis* placed side by side with *H. pusilla* appears very close even specifically.

DIBOLIA.

D. TEPPERI, sp.nov.

Ovalis; convexa; nitida; ferruginea (certo visu supra viridimicans); capite prothoraceque rufo-æneis; elytris femorumque posticorum apice fusco-æneis; prothorace subtiliter transversim strigoso; elytris duplo-punctulatis, haud striatis.

[Long. 2⅛, lat. 1⅔ lines.

The eyes are very large, and nearly meet on the summit of the head. The elytra are very finely and very closely punctulate (this puncturation only visible under a powerful lens) and also provided with a system of less fine and less close (though actually fine and close) puncturation; they have no trace of longitudinal striæ. The prothorax is across its base about twice and a-half again as wide as it is long down the middle, its sides are nearly straight, its base is slightly bisinuate.

Allied to *D. Duboulayi*, Baly (from Western Australia) but differing *inter alia* in its larger size, its wholly ferruginous antennæ but little infuscate towards the apex, and its non-striate elytra.

N. Territory of S. Australia; taken by Mr. J. P. Tepper.

OIDES.

O. TEPPERI, sp.nov.

Flava; antennis mandibulis tibiisque (basi excepta), tarsis totis, et abdomine plus minusve, piceis vel nigris; elytris parte posteriori macula elongata magna cyanea ornatis; capite postice longitudinaliter canaliculato, inter oculos transversim impresso, leviter obscure punctulato; prothorace quam longiori duplo latiori, antice et postice leviter transversim impresso, subtiliter sparsius punctulato; elytris subtiliter sat crebre punctulatis.

[Long. 3½-4, lat. 1½-2 lines.

The basal two joints of the antennæ are entirely flavous, the following two are more or less infuscate or piceous towards the apex, the rest black ; the fourth joint is a little longer than the third. The hind body is infuscate to a variable extent, in some specimens the infuscation being confined to the middle part of the basal two or three segments while in others it suffuses the whole of the ventral segments except the last, leaving, however, a flavous margin down either side.

N. Territory of S. Australia ; taken by Mr. J. P. Tepper.

O. SOROR, sp.nov.

Flava ; antennis mandibulis tibiisque (basi excepta), et tarsis totis piceis vel nigris; elytris singulis (marginibus suturali laterali apicalique exceptis) cyaneo-nigris ; capite postice longitudinaliter canaliculato, inter oculos transversim impresso, obscure subcrasse punctulato ; prothorace quam longiori minus duplo latiori, inæquali, crebre sat fortiter punctulato, antice et postice transversim, et alibi, impresso ; elytris crebre subtilius punctulatis.

[Long. 3½, lat. 2 lines.

The antennæ are coloured as those of *O. Tepperi ;* the third and fourth joints are of equal length, the second very evidently shorter. The blue-black colouring on the elytra occupies the whole surface except a narrow border running entirely round each of them except at the base where it is wanting.

Several species of *Oides* more or less resembling this insect have been described from Australia and elsewhere, from all of which the combination of characters mentioned above will, I think, distinguish it. Of Australian species it is no doubt nearest to *O. circumdata,* Baly, in which, however, the second joint of the antennæ is as long as the third, and the prothorax is *finely* punctulate ; *O. lætabile,* Clark, has the hind body black and the lateral yellow margin not reaching the apex.

N. Territory of S. Australia ; taken by Mr. J. P. Tepper.

O. SILPHOMORPHOIDES, sp.nov.

Flava, vel flavo-fusca ; antennis mandibulisque (basi excepta),
et elytris vitta lata submarginali (nec basin nec apicem attingente)
postice gradatim dilatata, piceis ; capite longitudinaliter subtiliter
canaliculato et inter oculos transversim impresso, minute coriaceo
et punctis majoribus obscuris sparsim impresso ; prothorace quam
longiori plus duplo latiori, antice et postice transversim (et
utrinque longitudinaliter) impresso, capiti similiter punctulato ;
elytris crebre minus subtiliter punctulatis.

[Long. 3-3¾, lat. 1⅔-2 lines.

The tarsi of this species are scarcely infuscate. Of the antennæ
joints 1 and 2 are testaceous, 3-6 increasingly stained with piceous,
the rest entirely piceous ; joint 2 is short, 3 and 4 equal. On the
prothorax the sublateral longitudinal impressions connect the ends
of the transverse impressions, so that an oblong transverse discal
space, is enclosed. The elytra, as compared with those of allied
species, are rather strongly punctured, the head and prothorax
exceptionally feebly. The insect bears a considerable superficial
resemblance to a *Silphomorpha*.

AULACOPHORA.

A. PALMERSTONI, sp.nov.

Supra testacea vel fulva, antennis (basi excepta) et labro infuscatis ; subtus (capite prothorace et abdominis apice fulvis exceptis)
nigra, dense sat longe albido-pubescens ; tibiis apice et tarsis vix
infuscatis ; capite vix evidenter punctulato ; prothorace quam
longiori vix dimidia parte latiori, medio fortiter transversim
sulcato, latera versus subfortiter punctulato ; elytris crebre subtiliter punctulatis.

♂. Antennarum articulo primo modice triangulariter dilatato,
abdominis segmento apicali trilobato, lobo intermedio oblongo-
quadrato, profunde concavo, apice emarginato.

[Long. 3-3½, lat. 1⅔ lines.

A furrow runs across the head from eye to eye which is much stronger in the female than in the male. From between the bases of the antennæ a smooth ridge runs down the middle of the clypeus nearly to its apex.

N. Territory of Australia; taken Mr. J. P. Tepper and others.

A. AUSTRALIS, sp.nov.

Sat nitida; capite prothorace scutello elytrisque flavis; his basi fascia lata suturam fere attingente et macula magna subapicali nigra instructis, apice ipso anguste piceo; subtus flava abdomine apicem versus et metasterno nigris; antennis (basi excepta) tibiis tarsisque infuscatis; capite vix evidenter, prothorace (medio transversim fortiter sulcato) latera versus crebrius fortius, elytris subtilius minus crebre punctulatis.

♂. Antennarum articulis 3° (leviter) 4° 5°que (valde) dilatatis; abdominis segmento apicali longitudinaliter 4-sulcato, inter sulcos interstitiis convexis.

The basal black spot (or fascia) on the elytra occupies the anterior quarter extending from the lateral margin almost to the suture, its hinder and inner edges being irregular in outline; the hinder black spot is scarcely smaller than the basal one, and almost touches the lateral margin, being well separated from the suture, with its front edge a little behind the middle of the elytron. The basal joint of the antennæ is moderately elongate, the second short, third about equal to 1st (in the male somewhat dilated), fourth slightly shorter than third (in the ♂ strongly dilated and accuminate at the extero-apical angle, fifth in male dilated as strongly as fourth than which it is much shorter,—in female similar to fourth and scarcely shorter,—the remaining joints gradually and slightly (in both sexes) increasing in length and decreasing in thickness.

I have met with this insect in various localities near Adelaide, and have received specimens from N. S. Wales (from Mr. Sloane). It appears to be a common species, but I cannot discover any description of it among the numerous described forms of the genus.

In some respects it agrees with the description of *A. cartereti*, but the antennæ of that species are said to be as long as the body, the hinder black mark on the elytra is said to be "at the extremity," and the underside and legs are said to be "yellow" without any parts thereof being excepted,—in none of which respects does the present species agree with the description.

AGELASTICA.

A. IMPURA, sp.nov.

Elongato-ovalis, postice vix ampliata ; rufo-fulva ; capite (antennas includente), abdominis segmentis (ultimo excepto) in medio, femoribus (anticis totis, intermediis basi ipsa excepta, posticis dimidia parte apicali), tibiis, tarsisque, nigris ; prothoracis disco infuscato ; elytris violaceo-cœruleis ; prothorace impunctato obscure bifoveolato ; elytris sat crebre punctulatis. [Long. 3, lat. 1¾ lines.

The antennæ are nearly as long as the body, rather robust, the 2nd joint short, the 3rd twice as long, the 4th and following joints much longer still.

N. Territory of S. Australia ; taken by Mr. J. P. Tepper.

A. MELANOCEPHALA, Baly.

I have received from several collectors in the Northern Territory specimens of an insect which appears to be this species. It is not quite clear from Mr. Baly's description, however, whether in the phrase "*capite nigro*" he includes the antennæ (the colour of which is not specially mentioned). Those of the species before me are black.

RUPILIA.

R. IMPRESSA, sp.nov.

Rufa vel ferruginea ; antennis tibiis tarsisque piceis vel piceo-nigris ; elytris cyaneis cupreo-iridescentibus ; nonnullis exemplis sutura rufa, nonnullis abdomine supra et subtus genubusque piceo notatis ; capite inæquali, postice longitudinaliter fortiter canaliculato, obscure sat crasse punctulato ; prothorace quam longiori

fere dimidia parte latiori, antice quam postice paullo latiori, inæquali, trans medium impresso, subtiliter nec crebre punctulato,
disco in medio fere lævigato, margine antico in medio vix evidenter
(postico sat fortiter) emarginato, lateribus fere rectis ; scutello sat
magno, fovea magna circulari impresso ; elytris crebre subtilius
punctulatis, disco sulco longitudinali lato impresso.

[Long. 4, lat. 2 lines.

I refer this insect to *Rupilia* with some hesitation on account
of the structure of its tibiæ and antennæ, the former being bi-
canaliculate externally with the interval between the channels
strongly costiform, and the latter being quite $\frac{2}{3}$ the length of the
body with the apical joints scarcely dilated and the shortest of
them (8-10) very decidedly longer than wide.   I do not know of
any characterized genus presenting these features, but as the specimens before me agree very well with *Rupilia* in other respects,
I do not think it necessary to give them a new generic name.
Cyclippa seems to want the external keel of the tibiæ and to differ
widely in the style of colour and markings, while the specimens
before me seem to resemble the described species of *Rupilia* in the
latter respects.

The colour of the elytra is peculiar being a rather dull blue
with a kind of iridescence which in certain lights makes them
appear reddish violet or coppery ; their sutural apex reaches to
about the base of the antepenultimate segment of the hind body
while (owing to the obliquity of the truncation of their apical
margin) the external apex is on a level with the base of the
penultimate segment (these measurements may not be quite exact
as all the specimens before me are much distorted).   Immediately
behind the base of each elytron and a little within the humeral
callus a wide longitudinal depression commences, and extends to
near the apex, appearing as though the whole substance of the
elytron were indented ; the limits of this depression are not
defined but it occupies the whole middle half of the organ.   I am
doubtful of the sex of the specimens before me.   The transverse
furrow across the middle of the prothorax is much more conspicuous in some examples than in others.

From *Rupilia ruficollis*, Clark, which this insect must resemble rather closely, it would seem to be distinguished *inter alia* by the uniform colour of the antennæ, the much finer puncturation of the elytra, by the depressions on those organs and by that on the scutellum.

N. Territory of S. Australia; taken by Prof. Tate and by Mr. J. P. Tepper.

MENIPPUS.

M. MACULICOLLIS, sp.nov.

Oblongus ; robustus ; undique pube aurea adpressa vestitus ; fuscus vel ferrugineus ; vertice in medio, prothorace ad latera antice et basi in medio, elytris latera apicemque versus, scutello, antennis, mandibulis apice, femoribus maculis nonnullis, tibiis, tarsis, et meso-metaque sternis ad latera, nigro-piceis ; capite prothoraceque confuse obscure (hoc antice latera versus distincte sat crebre), scutello elytrisque crebre subfortiter, punctulatis ; capite postice longitudinaliter canaliculato ; prothorace quam longiori duplo latiori, antice late fortiter transversim arcuatim sulcato, marginibus antico et postico leviter subangulatim emarginatis, lateribus (sulci transversi incisura) pone medium emarginatis ; antennis longitudine corporis dimidio æqualibus, sat validis.

[Long. 4½, lat. 2⅓ lines.

The characters of this insect seem to agree very well in all respects with those attributed to *Menippus*. The colour of the elytra varies a good deal, the ground tint in dark specimens being so pitchy as to obscure the markings ; in the darkest specimen before me the elytra are of an almost unicolorous pitchy black. The short golden pubescence with which the insect is clothed is spread over the whole surface including the legs and antennæ but seems to be very deciduous on the head and prothorax which in most of the examples before me are nitid and almost glabrous.

The black spots on the head and prothorax will distinguish this species from *M. cynicus*, Clark, also from *Galeruca semipullata*, Clk., which latter moreover seems to have simple claws since

Mr. Clark attributes it to *Galeruca* on the same page on which he distinguishes *Menippus* from that genus by its claws not being simple.

N. Territory of S. Australia; taken by Dr. Wood and Mr. J. P. Tepper.

MONOLEPTA.

M. TEPPERI, sp.nov.

Elongato-oblonga; sat parallela; fusco-testacea; antennis, tibiis tarsisque piceis; prothorace femoribusque flavo- (magis quam fusco-) testaceis; elytris disco longitudinaliter infuscatis, spatio infuscato nec basin nec apicem attingente; supra sat æqualiter subtilius crebre obscure punctulata.   [Long. 2⅔, lat. 1 line.

The head is transversely grooved behind the insertion of the antennæ between which an obscure keel takes its rise and runs forward for a short distance. The antennæ are unfortunately broken in both the specimens before me, but are probably a little more than half the length of the body; the basal joint is elongate and gently thickened towards the apex, its extreme base being testaceous, its remainder dark shining brown; joints 2-4 are dull pitchy black, 2 short, 3 longer, 4 longer still; the rest are wanting. The prothorax is subquadrate, a little more than a half wider than long, the sides but little rounded, the front subtruncate, not much narrower than the base which is rounded. The scutellum is triangular and rather small. The vitta-like infuscation on the disc of each elytron leaves a narrow lateral, and a wide sutural, pale margin. Only the apex of the pygidium is exposed. The basal joint of the posterior tarsi is slightly longer than the following three together; the posterior tibiæ are armed with a long spine; the anterior coxal cavities are closed; the elytral epipleuræ wide near the base and quite obscure beyond the middle.

Appears to be allied to *M. dimidiata*, Jacoby, (from Cape York) but differs from it structurally in the pygidium being almost covered by the elytra.

N. Territory of S. Australia; taken by Mr. J. P. Tepper.

EURISPA.

E. MAJOR, sp.nov.

Piceo-nigra; supra (capite obscuriore excepto) testacea, abdomine medio rufescenti; prothorace quam latiori tertiâ parte longiori, ante medium constricto, crasse profunde punctulato; elytris punctulato-striatis, vix evidenter quadricostatis, apice 'valde productis, spinosis; unguiculis nullis. [Long. 4½-5, lat. 1 line (vix).

The anterior constriction and large deep puncturation of the prothorax, and the sculpture of the elytra (which are punctulate-striate, with all the interstices subcostate,—4 of them slightly more strongly and widely than the others) will distinguish this species from all its previously described congeners.

N. Territory of S. Australia; taken by Mr. J. P. Tepper.

EROTYLIDÆ.

THALLIS.

The species of this genus seem to be rather more widely distributed than most of the Australian *Coleoptera*. I have found at Port Lincoln T. *janthina, compta* and *vinula* described by Erichson on Tasmanian specimens, also a species which does not appear to differ from *T. Erichseni*, Crotch (described from N. S. Wales), and another which I take to be *T. insueta*, Crotch (described from Rockhampton, Queensland). The insect last mentioned displays all the strongly marked characters which led Mr. Crotch to hesitate in referring *insueta* to *Thallis*, and is similarly coloured (though a little more brightly than the description would lead one to expect); it differs, however, in having the prosternum a little more prominent behind the coxæ and a little less coarsely sculptured in front than that of *T. insueta* is said to be, but I do not think it can be regarded as a distinct species,—at least without an actual comparison with the type.

EPISCAPHULA.

E. GUTTATIPENNIS, sp.nov.

Picea ; prothorace antice et ad latera, elytris singulis maculis 5 parvis, abdomine, et tarsis, rufis ; capite subtilius parcius, prothorace (præsertim latera versus) crebre sat fortiter, punctulatis ; elytris seriatim punctulatis, vix striatis, interstitiis crebrius minus subtiliter punctulatis ; subtus subtilius minus crebre punctulata.
[Long. 2¼, lat. 1 line (vix).

The red spots on each elytron are all small and are placed as follows :—three on the disc at a distance from the base of a quarter, two-thirds, and five-sixths the length of the elytra ; and two (much smaller than those on the disc) near the lateral margin, —one level with the first, and the other a little behind the second, of the discal spots. The reddish colour of the ventral segments is brighter down the middle than at the sides, and extends itself a little on the metasternum. The ventral segments bear some golden pubescence.

A much more parallel insect than the following, and appearing to hover between *Thallis* and *Episcaphula*, resembling the former in general appearance but having the elongate second antennal joint (about twice as long as the third joint) and the triangularly emarginate prosternum (receiving the pointed apex of the mesosternum) of the latter. The sides of the prosternum are strongly carinate. The prothorax is strongly transverse and very little narrowed anteriorly. The condition of the specimen before me, though in other respects very good, precludes any reliable descrip-of the mouth organs.

N. Territory of S. Australia ; taken by Mr. J. P. Tepper.

E. DUPLOPUNCTATA, sp.nov.

Rufescens ; capite, prothoracis ad basin macula quadrata, elytris (fasciis tribus suturam versus abbreviatis exceptis)

nigricantibus ; capite crebre fortiter, prothorace duplo (crebre
subtilius et latera versus crassissime sparsius), punctulatis ; elytris
vix striatis, striis seriatim punctulatis, interstitiis subtilius sat
crebre punctulatis ; subtus crebre fortius (metasterno subtiliter)
punctulata ; segmentis ventralibus pubescentibus.

[Long. 2⅘, lat. 1 line.

The ferruginous tone of the antennæ, legs, and metasternum is
considerably darker than that of the under surface in general.
The red fasciæ on the elytra are as follows : one near the base
wavy and about as wide as a sixth of the length of the elytra,
emitting from the middle of its front margin a projection which
almost touches the base ; a second considerably behind the middle,
narrower than the anterior one and gently curved forward ; a
third close to the apex equal in width to the second ; these fasciæ
all touch the lateral margins and nearly reach the suture. The
prothorax is strongly transverse and is evidently narrowed
anteriorly. The prosternum has its process margined and
triangularly emarginate at the apex. The greatest width of
the insect is near the front of the elytra whence it is much narrowed hindward.

Allied to *E. rudepunctata,* Crotch, but differing *inter alia* in the
coarse punctures of the prothorax being quite confined to the sides,
in the prothorax having a large quadrate dark spot at the base, in
the very different shape of the elytral fasciæ, and in the very
evident rather close puncturation of the elytral interstices.

N. Territory of S. Australia ; taken by Mr. J. P. Tepper.

CŒLOPHORA.

C. PUPILLATA, Muls.

Among some specimens forwarded to me from the N. Territory
of S. Australia by Dr. Bovill I find an example that evidently
pertains to this species, hitherto not noticed as Australian ; it is
known as inhabiting India, China, and Java. Several of its
congeners, though omitted in Mr. Masters' Catalogue, are
mentioned by Mulsant as occurring in Australia.

COCCINELLIDÆ.

CRYPTOLÆMUS.

C. MONTROUZIERI, Muls.

I have recently received from F. M. Bailey, Esq., of Brisbane, two specimens of this pretty little *Scymnid*, taken in the Brisbane neighbourhood. The habitat of the species as given by its author is "Australia" merely; it is omitted altogether in Mr. Masters' Catalogue.

C. SIMPLEX, sp.nov.

Late ovalis; pubescens; subtilius sat dense subæqualiter punctulatus; ferrugineus; elytris nigris, apice rufescentibus; metasterno et abdominis segmento basali in medio infuscatis.

[Long. 2½, lat. 1¾ lines.

Very like *C. Montrouzieri*, but a little wider,—especially behind,—and differently coloured, the legs and underside (except a slight infuscation of the latter) being entirely ferruginous, the scutellum red and the apex of the elytra more narrowly reddened.

Sent to me from the Northern Territory of S. Australia by Dr. Bovill.

www.ingramcontent.com/pod-product-compliance
Lightning Source LLC
Chambersburg PA
CBHW021822190326
41518CB00007B/708